Graduate Texts in Mathematics **90**

Editorial Board
F. W. Gehring P. R. Halmos (Managing Editor)
C. C. Moore

Arne Brøndsted

An Introduction to Convex Polytopes

Springer-Verlag
New York Heidelberg Berlin

Arne Brøndsted
Københavns Universitets
Matematiske Institut
Universitetsparken 5
2100 København Ø
Danmark

Editorial Board

P. R. Halmos
Managing Editor
Indiana University
Department of Mathematics
Bloomington, IN 47405
U.S.A.

F. W. Gehring
University of Michigan
Department of Mathematics
Ann Arbor, MI 48104
U.S.A.

C. C. Moore
University of California at Berkeley
Department of Mathematics
Berkeley, CA 94720
U.S.A.

AMS Subject Classifications (1980): 52-01, 52A25

Library of Congress Cataloging in Publication Data
Brøndsted, Arne.
An introduction to convex polytopes.
(Graduate texts in mathematics; 90)
Bibliography: p.
1. Convex polytopes. I. Title. II. Series.
QA64.0.3.B76 1982 514'.223 82-10585

With 3 Illustrations.

© 1983 by Springer-Verlag New York Inc.
All rights reserved. No part of this book may be translated or reproduced in any form without written permission from Springer-Verlag, 175 Fifth Avenue, New York, New York 10010, U.S.A.

Typeset by Composition House Ltd., Salisbury, England.
Printed and bound by R. R. Donnelley & Sons, Harrisonburg, VA.
Printed in the United States of America.

9 8 7 6 5 4 3 2 1

ISBN 0-387-90722-X Springer-Verlag New York Heidelberg Berlin
ISBN 3-540-90722-X Springer-Verlag Berlin Heidelberg New York

Preface

The aim of this book is to introduce the reader to the fascinating world of convex polytopes.

The highlights of the book are three main theorems in the combinatorial theory of convex polytopes, known as the Dehn–Sommerville Relations, the Upper Bound Theorem and the Lower Bound Theorem. All the background information on convex sets and convex polytopes which is needed to understand and appreciate these three theorems is developed in detail. This background material also forms a basis for studying other aspects of polytope theory.

The Dehn–Sommerville Relations are classical, whereas the proofs of the Upper Bound Theorem and the Lower Bound Theorem are of more recent date: they were found in the early 1970's by P. McMullen and D. Barnette, respectively. A famous conjecture of P. McMullen on the characterization of f-vectors of simplicial or simple polytopes dates from the same period; the book ends with a brief discussion of this conjecture and some of its relations to the Dehn–Sommerville Relations, the Upper Bound Theorem and the Lower Bound Theorem. However, the recent proofs that McMullen's conditions are both sufficient (L. J. Billera and C. W. Lee, 1980) and necessary (R. P. Stanley, 1980) go beyond the scope of the book.

Prerequisites for reading the book are modest: standard linear algebra and elementary point set topology in $\mathbb{R}^d$ will suffice.

The author is grateful to the many people who have contributed to the book: several colleagues, in particular Victor Klee and Erik Sparre Andersen, supplied valuable information; Aage Bondesen suggested essential improvements; students at the University of Copenhagen also suggested improvements; and Ulla Jacobsen performed an excellent typing job.

Copenhagen
February 1982

ARNE BRØNDSTED

Contents

Introduction

Convex polytopes are the d-dimensional analogues of 2-dimensional convex polygons and 3-dimensional convex polyhedra. The theme of this book is the combinatorial theory of convex polytopes. Generally speaking, the combinatorial theory deals with the numbers of faces of various dimensions (vertices, edges, etc.). An example is the famous theorem of Euler, which states that for a 3-dimensional convex polytope, the number f_0 of vertices, the number f_1 of edges and the number f_2 of facets are connected by the relation

$$f_0 - f_1 + f_2 = 2.$$

(In contrast to the combinatorial theory, there is a metric theory, dealing with such notions as length, angles and volume. For example, the concept of a regular polytope belongs to the metric theory.)

The main text is divided into three chapters, followed by three appendices. The appendices supply the necessary background information on lattices, graphs and combinatorial identities. Following the appendices, and preceding the bibliography, there is a section with bibliographical comments. Each of Sections 1–15 ends with a selection of exercises.

Chapter 1 (Sections 1–6), entitled "Convex Sets," contains those parts of the general theory of d-dimensional convex sets that are needed in what follows. Among the basic notions are the convex hull, the relative interior of a convex set, supporting hyperplanes, faces of closed convex sets and polarity. (Among the basic notions of convexity theory not touched upon we mention convex cones and convex functions.)

The heading of Chapter 2 (Sections 7–15) is "Convex Polytopes." In Sections 7–11 we apply the general theory of convex sets developed in Chapter 1 to the particular case of convex polytopes. (It is the author's belief that many properties of convex polytopes are only appreciated when seen on the background of properties of convex sets in general.) In Sections 12–14 the important classes of simple, simplicial, cyclic and neighbourly polytopes are introduced. In Section 15 we study the graph determined by the vertices and edges of a polytope.

Chapter 3 contains selected topics in the "Combinatorial Theory of Convex Polytopes." We begin, in Section 16, with Euler's Relation in its d-dimensional version. In Section 17 we discuss the so-called Dehn–Sommerville Relations which are "Euler-type" relations, valid for simple or simplicial polytopes only. Sections 18 and 19 are devoted to the celebrated Upper Bound Theorem and Lower Bound Theorem, respectively; these theorems solve important extremum problems involving the numbers of faces (of various dimensions) of simple or simplicial polytopes. Finally, in Section 20 we report on a recent fundamental theorem which gives "complete information" on the numbers of faces (of various dimensions) of a simple or simplicial polytope.

The following flow chart outlines the organization of the book. However, there are short cuts to the three main theorems of Chapter 3. To read the proof of the Dehn–Sommerville Relations (Theorem 17.1) only Sections 1–12 and Euler's Relation (Theorem 16.1) are needed; Euler's Relation also requires Theorem 15.1. To read the proof of the Upper Bound Theorem (Theorem 18.1) only Sections 1–14 and Theorems 15.1–15.3 are needed. To read the Lower Bound Theorem (Theorem 19.1) only Sections 1–12 and 15, and hence also Appendix 2, are needed. It is worth emphasizing that none of the three short cuts requires the somewhat technical Appendix 3.

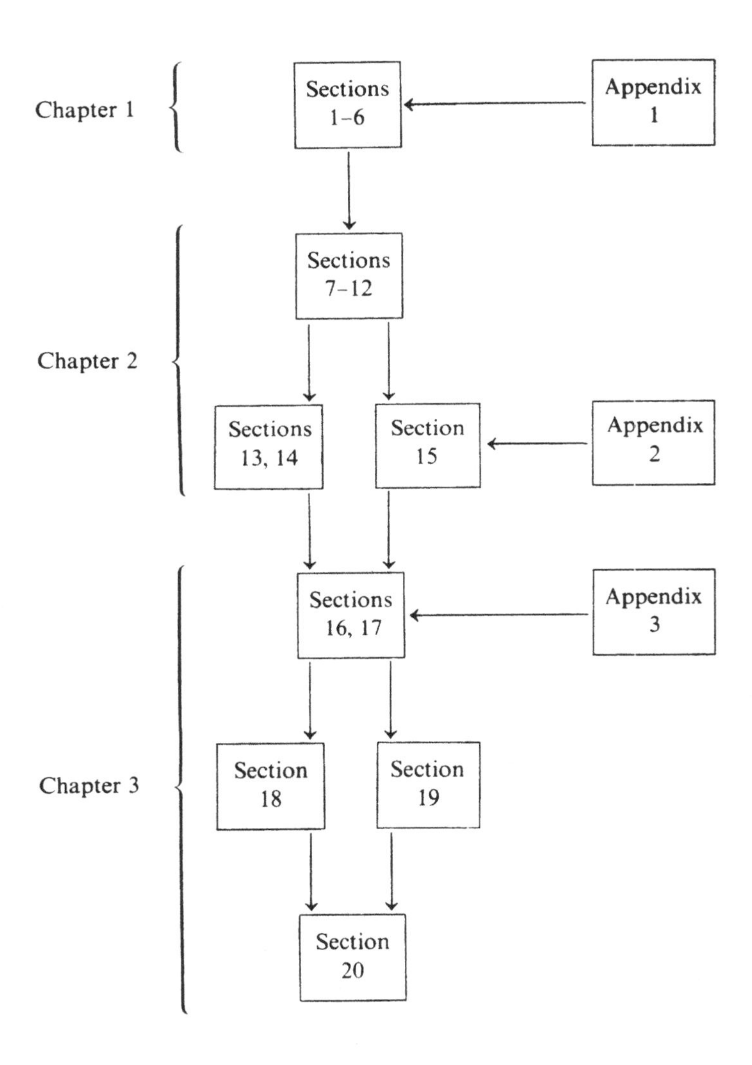
Chapter 1
Sections 1–6
Appendix 1
Chapter 2
Sections 7–12
Sections 13, 14
Section 15
Appendix 2
Chapter 3
Sections 16, 17
Appendix 3
Section 18
Section 19
Section 20

CHAPTER 1

Convex Sets

§1. The Affine Structure of $\mathbb{R}^d$

The theory of convex polytopes, and more generally the theory of convex sets, belongs to the subject of affine geometry. In a sense, the right framework for studying convex sets is the notion of a Euclidean space, i.e. a finite-dimensional real affine space whose underlying linear space is equipped with an inner product. However, there is no essential loss of generality in working only with the more concrete spaces $\mathbb{R}^d$; therefore, everything will take place in $\mathbb{R}^d$. We will assume that the reader is familiar with the standard linear theory of $\mathbb{R}^d$, including such notions as subspaces, linear independence, dimension, and linear mappings. We also assume familiarity with the standard inner product $\langle \cdot, \cdot \rangle$ of $\mathbb{R}^d$, including the induced norm $\|\cdot\|$, and elementary topological notions such as the interior int M, the closure cl M, and the boundary bd M of a subset M of $\mathbb{R}^d$.

The main purpose of this section is to give a brief survey of the affine structure of $\mathbb{R}^d$. We give no proofs here; the reader is invited to produce his own proofs, essentially by reducing the statements in the affine theory to statements in the linear theory. It is important that the reader feels at home in the affine structure of $\mathbb{R}^d$.

For $d \in \mathbb{N}$, we denote by $\mathbb{R}^d$ the set of all d-tuples $x = (\alpha_1, \ldots, \alpha_d)$ of real numbers $\alpha_1, \ldots, \alpha_d$. We identify $\mathbb{R}^1$ with $\mathbb{R}$, and we define $\mathbb{R}^0 := \{0\}$.

We recall some basic facts about the *linear structure* of $\mathbb{R}^d$. Equipped with the standard linear operations, $\mathbb{R}^d$ is a *linear space*. When the linear structure of $\mathbb{R}^d$ is in the foreground, the elements of $\mathbb{R}^d$ are called *vectors*. The zero vector is denoted by o.

A *linear subspace* is a non-empty subset L of $\mathbb{R}^d$ such that

(a) *$\lambda_1 x_1 + \lambda_2 x_2$ is in L for all $x_1, x_2 \in L$ and all $\lambda_1, \lambda_2 \in \mathbb{R}$.*

A *linear combination* of vectors $x_1, \ldots, x_n$ from $\mathbb{R}^d$ is a vector of the form $\lambda_1 x_1 + \cdots + \lambda_n x_n$, where $\lambda_1, \ldots, \lambda_n$ are in $\mathbb{R}$. Corresponding to $n = 0$, we allow the empty linear combination with the value o. (In the definition of a linear combination there is a certain ambiguity. In some situations when talking about a linear combination $\lambda_1 x_1 + \cdots + \lambda_n x_n$ we not only think of the vector $x = \lambda_1 x_1 + \cdots + \lambda_n x_n$, but also of the particular coefficients $\lambda_1, \ldots, \lambda_n$ used to represent x.) The condition (a) expresses that any linear combination of two vectors from L is again in L. Actually, (a) is equivalent to the following:

(b) *Any linear combination of vectors from L is again in L.*

(Strictly speaking, (a) and (b) are only equivalent when $L \neq \emptyset$. For $L = \emptyset$, condition (a) holds, whereas (b) is violated by the fact that we allow the empty linear combination. Note, however, that we did require $L \neq \emptyset$ in the definition of a linear subspace.)

The intersection of any family of linear subspaces of $\mathbb{R}^d$ is again a linear subspace of $\mathbb{R}^d$. Therefore, for any subset M of $\mathbb{R}^d$ there is a smallest linear subspace containing M, namely, the intersection of all linear subspaces containing M. This subspace is called the linear subspace *spanned* by M, or the *linear hull* of M, and is denoted by span M.

One has the following description of the linear hull of a subset M:

(c) *For any subset M of $\mathbb{R}^d$, the linear hull* span M *is the set of all linear combinations of vectors from M.*

(Note that our convention concerning the empty linear combination ensures that the correct statement span $\emptyset = \{o\}$ is included in (c).)

An n-family $(x_1, \ldots, x_n)$ of vectors from $\mathbb{R}^d$ is said to be *linearly independent* if a linear combination $\lambda_1 x_1 + \cdots + \lambda_n x_n$ can only have the value o when $\lambda_1 = \cdots = \lambda_n = 0$. (Note that the empty family, corresponding to $n = 0$, is linearly independent.) Linear independence is equivalent to saying that none of the vectors is a linear combination of the remaining ones. When a vector x is a linear combination of $x_1, \ldots, x_n$, say $x = \lambda_1 x_1 + \cdots + \lambda_n x_n$, then the coefficients $\lambda_1, \ldots, \lambda_n$ are uniquely determined if and only if $(x_1, \ldots, x_n)$ is linearly independent. An n-family $(x_1, \ldots, x_n)$ which is not linearly independent is said to be *linearly dependent*.

A *linear basis* of a linear subspace L of $\mathbb{R}^d$ is a linearly independent n-family $(x_1, \ldots, x_n)$ of vectors from L such that $L = \text{span}\{x_1, \ldots, x_n\}$. The *dimension* dim L of L is the largest non-negative integer n such that some n-family of vectors from L is linearly independent. A linearly independent n-family of vectors from L is a basis of L if and only if $n = \dim L$.

Let M be any subset of $\mathbb{R}^d$, and let n be the dimension of span M. Then there is actually a linearly independent n-family $(x_1, \ldots, x_n)$ of vectors from M, i.e. there is a basis $(x_1, \ldots, x_n)$ of span M consisting of vectors from M. We therefore have:

(d) *For any subset M of $\mathbb{R}^d$, there exists a linearly independent family $(x_1, \ldots, x_n)$ of vectors from M such that* span M *is the set of all linear combinations*

$$\sum_{i=1}^{n} \lambda_i x_i$$

of $x_1, \ldots, x_n$.

This statement is a sharpening of (c). It shows that to generate span M we need only take all linear combinations of the *fixed* vectors $x_1, \ldots, x_n$ from M. Furthermore, each vector in span M has a unique representation as a linear combination of $x_1, \ldots, x_n$.

A mapping φ from some linear subspace L of $\mathbb{R}^d$ into $\mathbb{R}^e$ is called a *linear mapping* if it preserves linear combinations, i.e.

$$\varphi\left(\sum_{i=1}^{n} \lambda_i x_i\right) = \sum_{i=1}^{n} \lambda_i \varphi(x_i).$$

When φ is linear, then $\varphi(L)$ is a linear subspace of $\mathbb{R}^e$. Linear mappings are continuous.

A one-to-one linear mapping from a linear subspace L_1 of $\mathbb{R}^d$ onto a linear subspace L_2 of $\mathbb{R}^e$ is called a (linear) *isomorphism.* If there exists an isomorphism from L_1 onto L_2, then L_1 and L_2 are said to be *isomorphic.* Two linear subspaces are isomorphic if and only if they have the same dimension. An isomorphism is also a homeomorphism, i.e. it preserves the topological structure.

We next move on to a discussion of the *affine structure* of $\mathbb{R}^d$. An *affine subspace* of $\mathbb{R}^d$ is either the empty set $\varnothing$ or a translate of a linear subspace, i.e. a subset $A = x + L$ where $x \in \mathbb{R}^d$ and L is a linear subspace of $\mathbb{R}^d$. (Note that L is unique whereas x can be chosen arbitrarily in A.) By an *affine space* we mean an affine subspace of some $\mathbb{R}^d$. When A_1 and A_2 are affine subspaces of $\mathbb{R}^d$ with $A_1 \subset A_2$, we shall also call A_1 an affine subspace of A_2. The elements $x = (\alpha_1, \ldots, \alpha_d)$ of some affine subspace A of $\mathbb{R}^d$ will be called *points* when the affine structure, rather than the linear structure, is in the foreground. (However, it will not always be possible, nor desirable, to distinguish between points and vectors.)

A subset A of $\mathbb{R}^d$ is an affine subspace if and only if the following holds:

(a′) *$\lambda_1 x_1 + \lambda_2 x_2$ is in A for all $x_1, x_2 \in A$ and all $\lambda_1, \lambda_2 \in \mathbb{R}$ with $\lambda_1 + \lambda_2 = 1$.*

For any two distinct points x_1 and x_2 in $\mathbb{R}^d$, the set

$$\{\lambda_1 x_1 + \lambda_2 x_2 \mid \lambda_1, \lambda_2 \in \mathbb{R}, \lambda_1 + \lambda_2 = 1\}$$

is called the *line* through x_1 and x_2. The condition (a′) then states that the line through any two points of A is contained in A.

An *affine combination* of points $x_1, \ldots, x_n$ from $\mathbb{R}^d$ is a linear combination $\lambda_1 x_1 + \cdots + \lambda_n x_n$, where $\lambda_1 + \cdots + \lambda_n = 1$. We shall write

$$\sum_{i=1}^{n}{}^{a} \lambda_i x_i$$

to indicate that the linear combination $\lambda_1 x_1 + \cdots + \lambda_n x_n$ is in fact an affine combination. (The empty linear combination is not an affine combination. Therefore, in an affine combination $\lambda_1 x_1 + \cdots + \lambda_n x_n$ we always have $n \geq 1$.) The condition (a′) states that any affine combination of two points from A is again in A. Actually, (a′) is equivalent to the following:

(b′) *Any affine combination of points from A is again in A.*

The intersection of any family of affine subspaces of $\mathbb{R}^d$ is again an affine subspace of $\mathbb{R}^d$. (Here it is important to note that $\emptyset$ is an affine subspace.) Therefore, for any subset M of $\mathbb{R}^d$ there is a smallest affine subspace containing M, namely, the intersection of all affine subspaces containing M. This affine subspace is called the affine subspace *spanned* by M, or the *affine hull* of M, and it is denoted by aff M.

One has the following description of the affine hull of a subset M:

(c′) *For any subset M of $\mathbb{R}^d$, the affine hull* aff M *is the set of all affine combinations of points from M.*

An n-family $(x_1, \ldots, x_n)$ of points from $\mathbb{R}^d$ is said to be *affinely independent* if a linear combination $\lambda_1 x_1 + \cdots + \lambda_n x_n$ with $\lambda_1 + \cdots + \lambda_n = 0$ can only have the value o when $\lambda_1 = \cdots = \lambda_n = 0$. (In particular, the empty family, corresponding to $n = 0$, is affinely independent.) Affine independence is equivalent to saying that none of the points is an affine combination of the remaining points. When a point x is an affine combination of $x_1, \ldots, x_n$, say $x = \lambda_1 x_1 + \cdots + \lambda_n x_n$, then the coefficients $\lambda_1, \ldots, \lambda_n$ are uniquely determined if and only if $(x_1, \ldots, x_n)$ is affinely independent. An n-family $(x_1, \ldots, x_n)$ which is not affinely independent is said to be *affinely dependent.*

Affine independence of an n-family $(x_1, \ldots, x_n)$ is equivalent to linear independence of one/all of the $(n-1)$-families

$$(x_1 - x_i, \ldots, x_{i-1} - x_i, x_{i+1} - x_i, \ldots, x_n - x_i), \qquad i = 1, \ldots, n.$$

An *affine basis* of an affine space A is an affinely independent n-family $(x_1, \ldots, x_n)$ of points from A such that $A = \operatorname{aff}\{x_1, \ldots, x_n\}$. The *dimension* dim A of a non-empty affine space A is the dimension of the linear subspace L such that $A = x + L$. (Since L is unique, dim A is well defined. When A is a linear subspace, then the affine dimension and the linear dimension are the same by definition, and therefore we may use the same notation.) When $A = \emptyset$, we put dim $A = -1$. The dimension of A is then $n - 1$ if and only if n is the largest non-negative integer such that there is an affinely independent

n-family of points from A. An affinely independent n-family of points from A is an affine basis of A if and only if $n = \dim A + 1$.

Let M be any subset of $\mathbb{R}^d$, and let the dimension of aff M be $n - 1$. Then there is actually an affinely independent n-family $(x_1, \ldots, x_n)$ of points from M, i.e. there is an affine basis $(x_1, \ldots, x_n)$ of aff M consisting of points from M. We therefore have:

(d') *For any subset M of $\mathbb{R}^d$, there exists an affinely independent family $(x_1, \ldots, x_n)$ of points from M such that* aff M *is the set of all affine combinations*

$$\sum_{i=1}^{n}{}^{a} \lambda_i x_i$$

of $x_1, \ldots, x_n$.

This statement is a sharpening of (c'). It shows that to generate aff M it suffices to take all affine combinations of the *fixed* points $x_1, \ldots, x_n$ from M. Furthermore, each point in aff M has a unique representation as an affine combination of $x_1, \ldots, x_n$.

The 0-dimensional affine spaces are the 1-point sets. The 1-dimensional affine spaces are called *lines*. When x_1 and x_2 are two distinct points of $\mathbb{R}^d$, then the 2-family (x_1, x_2) is affinely independent. Therefore, aff$\{x_1, x_2\}$ is 1-dimensional, i.e. a line, and it is in fact the line through x_1 and x_2 in the sense used earlier in this section. Conversely, the line through two points x_1 and x_2 in the earlier sense is in fact a 1-dimensional affine space, i.e. a line.

An $(n - 1)$-dimensional affine subspace of an n-dimensional affine space A, where $n \geq 1$, is called a *hyperplane* in A. If A is an affine subspace of $\mathbb{R}^d$, then the hyperplanes in A are the sets $H \cap A$ where H is a hyperplane in $\mathbb{R}^d$ such that $H \cap A$ is a non-empty proper subset of A.

A mapping φ from an affine subspace A of $\mathbb{R}^d$ into $\mathbb{R}^e$ is called an *affine mapping* if it preserves affine combinations, i.e.

$$\varphi\left(\sum_{i=1}^{n}{}^{a} \lambda_i x_i\right) = \sum_{i=1}^{n}{}^{a} \lambda_i \varphi(x_i).$$

When φ is affine, then $\varphi(A)$ is an affine subspace of $\mathbb{R}^e$. When $A = x + L$, where L is a linear subspace of $\mathbb{R}^d$, then a mapping $\varphi: A \to \mathbb{R}^e$ is affine if and only if there exists a linear mapping $\Phi: L \to \mathbb{R}^e$ and a point $y \in \mathbb{R}^e$ such that $\varphi(x + z) = y + \Phi(z)$ for all $z \in L$. Affine mappings are continuous.

An affine mapping $\varphi: A \to \mathbb{R}$ is called an *affine function* on A. For each hyperplane H in A there is a (non-constant) affine function φ on A such that $H = \varphi^{-1}(0)$. Conversely, $\varphi^{-1}(0)$ is a hyperplane in A for each non-constant affine function φ on A. We have $\varphi^{-1}(0) = \psi^{-1}(0)$ for two affine functions φ and ψ on A if and only if $\varphi = \lambda\psi$ for some non-zero real λ.

When φ is a non-constant affine function on an affine space A, we shall call the sets $\varphi^{-1}(]-\infty, 0[)$ and $\varphi^{-1}(]0, +\infty[)$ the *open halfspaces* bounded by the hyperplane $H = \varphi^{-1}(0)$, and we shall call the sets $\varphi^{-1}(]-\infty, 0])$ and $\varphi^{-1}([0, +\infty[)$ the *closed halfspaces* bounded by $H = \varphi^{-1}(0)$. Open halfspaces are non-empty open sets, closed halfspaces are non-empty closed sets. If $H = \varphi^{-1}(0)$ is a hyperplane in A, then two points from $A \setminus H$ are said to be on the *same side* of H if they both belong to $\varphi^{-1}(]-\infty, 0[)$ or both belong to $\varphi^{-1}(]0, +\infty[)$; if each of the two open halfspaces contains one of the two points, we shall say that they are on *opposite sides* of H.

A *halfline* is a halfspace in a line.

Let A be an affine subspace of $\mathbb{R}^d$, and let K be a closed halfspace in $\mathbb{R}^d$ such that $A \cap K$ is a non-empty proper subset of A. Then $A \cap K$ is a closed halfspace in A. Conversely, each closed halfspace in A arises this way. The same applies to open halfspaces.

For $y \in \mathbb{R}^d$ and $\alpha \in \mathbb{R}$ we let

$$H(y, \alpha) := \{x \in \mathbb{R}^d \mid \langle x, y \rangle = \alpha\}.$$

Note that $H(o, \alpha) = \mathbb{R}^d$ when $\alpha = 0$, and $H(o, \alpha) = \emptyset$ when $\alpha \neq 0$. The fact that the affine functions on $\mathbb{R}^d$ are precisely the functions

$$x \mapsto \langle x, y \rangle - \alpha, \qquad y \in \mathbb{R}^d, \alpha \in \mathbb{R},$$

implies that the hyperplanes in $\mathbb{R}^d$ are precisely the sets $H(y, \alpha)$ for $y \neq o$. If $y \neq o$, then y is called a *normal* of $H(y, \alpha)$.

For $y \in \mathbb{R}^d$ and $\alpha \in \mathbb{R}$ we let

$$K(y, \alpha) := \{x \in \mathbb{R}^d \mid \langle x, y \rangle \leq \alpha\}.$$

Note that $K(o, \alpha) = \mathbb{R}^d$ when $\alpha \geq 0$, and $K(o, \alpha) = \emptyset$ when $\alpha < 0$. For $y \neq o$, the set $K(y, \alpha)$ is one of the two closed halfspaces in $\mathbb{R}^d$ bounded by $H(y, \alpha)$. The other closed halfspace bounded by $H(y, \alpha)$ is $K(-y, -\alpha)$. Note that

$$\begin{aligned} \text{bd}\, K(y, \alpha) &= H(y, \alpha), \\ \text{int}\, K(y, \alpha) &= K(y, \alpha) \setminus H(y, \alpha), \\ \text{cl}(\text{int}\, K(y, \alpha)) &= K(y, \alpha), \end{aligned}$$

when $y \neq o$.

A one-to-one affine mapping from an affine space A_1 onto an affine space A_2 is called an (affine) *isomorphism*. If there exists an isomorphism from A_1 onto A_2, then A_1 and A_2 are said to be (affinely) *isomorphic*. Two affine spaces are isomorphic if and only if they have the same dimension. An isomorphism is also a homeomorphism, i.e. it preserves the topological structure.

From what has been said above, it follows that any d-dimensional affine space A is affinely isomorphic to the particular d-dimensional affine space $\mathbb{R}^d$. In other words, A may be "identified" with $\mathbb{R}^d$, not only in an affine sense but also in a topological sense. Note also that any given point of A can be "identified" with any given point of $\mathbb{R}^d$.

Finally, we should like to point out that this section does not include all the necessary information about the linear and affine structure of $\mathbb{R}^d$ needed in what follows. Some important additional information is contained in Exercises 1.1–1.5.

Exercises

1.1. Let $(x_1, \ldots, x_n)$ be an n-family of points from $\mathbb{R}^d$, where

$$x_i = (\alpha_{1i}, \ldots, \alpha_{di}), \qquad i = 1, \ldots, n.$$

Let

$$\bar{x}_i := (1, \alpha_{1i}, \ldots, \alpha_{di}), \qquad i = 1, \ldots, n.$$

Show that the n-family $(x_1, \ldots, x_n)$ is affinely independent if and only if the n-family $(\bar{x}_1, \ldots, \bar{x}_n)$ of vectors from $\mathbb{R}^{d+1}$ is linearly independent.

1.2. For any subset M of $\mathbb{R}^d$, show that

$$\dim(\operatorname{aff} M) = \dim(\operatorname{span} M)$$

when $o \in \operatorname{aff} M$, and

$$\dim(\operatorname{aff} M) = \dim(\operatorname{span} M) - 1$$

when $o \notin \operatorname{aff} M$.

1.3. Let A be an affine subspace of $\mathbb{R}^d$, and let H be a hyperplane in $\mathbb{R}^d$. Show that

$$\dim(A \cap H) = \dim A - 1$$

when $A \cap H \neq \emptyset$ and $A \not\subset H$.

1.4. Let $A_1 = x_1 + L_1$ and $A_2 = x_2 + L_2$ be non-empty affine subspaces of $\mathbb{R}^d$. Then A_1 and A_2 are said to be *parallel* if $L_1 \subset L_2$ or $L_2 \subset L_1$, *complementary* if L_1 and L_2 are complementary (and *orthogonal* if L_1 and L_2 are orthogonal).

Show that if A_1 and A_2 are parallel and $A_1 \cap A_2 \neq \emptyset$, then $A_1 \subset A_2$ or $A_2 \subset A_1$.

Show that if A_1 and A_2 are complementary, then $A_1 \cap A_2$ is a 1-point set.

1.5. Let $A_1 = x_1 + L_1$ and $A_2 = x_2 + L_2$ be complementary affine subspaces of $\mathbb{R}^d$, and let x_0 be the unique common point of A_1 and A_2, cf. Exercise 1.4. Then $A_1 = x_0 + L_1$ and $A_2 = x_0 + L_2$. Let $\Pi\colon \mathbb{R}^d \to L_1$ denote the projection in the direction of L_2. For any $x \in \mathbb{R}^d$, let $\pi(x) := x_0 + \Pi(x - x_0)$. Show that $\pi(x)$ is the unique common point of A_1 and $(x - x_0) + A_2$. (The mapping π is called the *projection* onto A_1 in the direction of A_2. When A_1 and A_2 are also orthogonal, then π is called the *orthogonal projection* onto A_1.)

1.6. An n-family $(x_1, \ldots, x_n)$ of points from $\mathbb{R}^d$ is said to be in *general position* if every subfamily $(x_{i_1}, \ldots, x_{i_p})$ with $p \leq d + 1$ is affinely independent. Verify that $(x_1, \ldots, x_n)$ is in general position if and only if for each k with $0 \leq k \leq d - 1$ and for each k-dimensional affine subspace A of $\mathbb{R}^d$, the number of i's such that $x_i \in A$ is at most $k + 1$.

1.7. Let $x_1, \ldots, x_n$ be distinct points in $\mathbb{R}^d$. Show that there is $w \neq o$ such that for each $\alpha \in \mathbb{R}$, the hyperplane $H(w, \alpha)$ contains at most one of the points $x_1, \ldots, x_n$.

§2. Convex Sets

In this section we shall introduce the notion of a convex set and we shall prove some basic facts about such sets. In Section 1 we demonstrated a strong analogy between linear concepts and affine concepts. This analogy carries over to convex concepts, though not in a complete fashion.

A subset C of $\mathbb{R}^d$ is called a *convex set* if $\lambda_1 x_1 + \lambda_2 x_2$ belongs to C for all $x_1, x_2 \in C$ and all $\lambda_1, \lambda_2 \in \mathbb{R}$ with $\lambda_1 + \lambda_2 = 1$ and $\lambda_1, \lambda_2 \geq 0$.

When x_1 and x_2 are distinct points from $\mathbb{R}^d$, then the set

$$\begin{aligned}[x_1, x_2] &:= \{\lambda_1 x_1 + \lambda_2 x_2 \mid \lambda_1, \lambda_2 \geq 0, \lambda_1 + \lambda_2 = 1\} \\ &= \{(1 - \lambda)x_1 + \lambda x_2 \mid \lambda \in [0, 1]\}\end{aligned}$$

is called the *closed segment* between x_1 and x_2. *Half-open segments* $]x_1, x_2]$, $[x_1, x_2[$ and *open segments* $]x_1, x_2[$ are defined analogously. With this notation, a set C is convex if and only if the closed segment between any two points of C is contained in C.

The affine subspaces of $\mathbb{R}^d$, including $\mathbb{R}^d$ and $\emptyset$, are convex. Any (closed or open) halfspace is convex.

The image of a convex set under an affine mapping is gain convex. In particular, translates of convex sets are again convex.

By a *convex combination* of points $x_1, \ldots, x_n$ from $\mathbb{R}^d$ we mean a linear combination $\lambda_1 x_1 + \cdots + \lambda_n x_n$, where $\lambda_1 + \cdots + \lambda_n = 1$ and $\lambda_1, \ldots, \lambda_n \geq 0$. Every convex combination is also an affine combination. We shall write

$$\sum_{i=1}^{n}{}^{c}\, \lambda_i x_i$$

to indicate that the linear combination $\lambda_1 x_1 + \cdots + \lambda_n x_n$ is in fact a convex combination. The definition of a convex set expresses that any convex combination of two points from the set is again in the set. We actually have:

Theorem 2.1. *A subset C of $\mathbb{R}^d$ is convex if and only if any convex combination of points from C is again in C.*

PROOF. If any convex combination of points from C is again in C, then, in particular, any convex combination of two points from C is in C. Therefore, C is convex.

Conversely, assume that C is convex. We shall prove by induction on n that any point from $\mathbb{R}^d$ which is a convex combination of n points from C is again in C. For $n = 1$ this is trivial, and for $n = 2$ it follows by definition. So, let n be at least 3, assume that any convex combination of fewer than n points from C is in C, and let

$$x = \sum_{i=1}^{n}{}^{c}\, \lambda_i x_i$$

be a convex combination of n points $x_1, \ldots, x_n$ from C. If $\lambda_i = 0$ for some i, then x is in fact a convex combination of fewer than n points from C, and so x belongs to C by hypothesis. If $\lambda_i \neq 0$ for all i, then $\lambda_i < 1$ for all i, whence, in particular, $1 - \lambda_1 > 0$. Therefore, we may write

$$\begin{aligned} x &= \sum_{i=1}^{n}{}^{c} \lambda_i x_i \\ &= \lambda_1 x_1 + \sum_{i=2}^{n} \lambda_i x_i \\ &= \lambda_1 x_1 + (1 - \lambda_1) \sum_{i=2}^{n} \frac{\lambda_i}{1 - \lambda_1} x_i. \end{aligned}$$

Here

$$y := \sum_{i=2}^{n} \frac{\lambda_i}{1 - \lambda_1} x_i$$

is in fact a convex combination since $\lambda_2 + \cdots + \lambda_n = 1 - \lambda_1$, and so y is in C by hypothesis. By the convexity of C then $\lambda_1 x_1 + (1 - \lambda_1)y$ is also in C, i.e. x is in C. □

It is clear that the intersection of any family of convex sets in $\mathbb{R}^d$ is again convex. Therefore, for any subset M of $\mathbb{R}^d$ there is a smallest convex set containing M, namely, the intersection of all convex sets in $\mathbb{R}^d$ containing M. This convex set is called the convex set *spanned* by M, or the *convex hull* of M, and it is denoted by $\operatorname{conv} M$.

It is clear that $\operatorname{conv}(x + M) = x + \operatorname{conv} M$ for any point x and any set M. More generally, it follows from Theorem 2.2 below that $\operatorname{conv}(\varphi(M)) = \varphi(\operatorname{conv} M)$ when φ is an affine mapping.

We have the following description of the convex hull of a set:

Theorem 2.2. *For any subset M of $\mathbb{R}^d$, the convex hull* $\operatorname{conv} M$ *is the set of all convex combinations of points from M.*

Proof. Let C denote the set of all convex combinations of points from M. Since $M \subset \operatorname{conv} M$, each $x \in C$ is also a convex combination of points from the convex set $\operatorname{conv} M$; the "only if" part of Theorem 2.1 then shows that $C \subset \operatorname{conv} M$. To prove the opposite inclusion, it suffices to show that C is a convex set containing M. Since each $x \in M$ has the trivial representation $x = 1x$ as a convex combination of points from M, it follows that $M \subset C$. To see that $\lambda_1 x_1 + \lambda_2 x_2$ is in C for each $x_1, x_2 \in C$ and each $\lambda_1, \lambda_2 \geq 0$ with $\lambda_1 + \lambda_2 = 1$, note that by definition x_1 and x_2 are convex combinations of points from M, say

$$x_1 = \sum_{i=1}^{n}{}^{c} \mu_i y_i, \qquad x_2 = \sum_{i=n+1}^{m}{}^{c} \mu_i y_i.$$

But then

$$\lambda_1 x_1 + \lambda_2 x_2 = \sum_{i=1}^{n} \lambda_1 \mu_i y_i + \sum_{i=n+1}^{m} \lambda_2 \mu_i y_i,$$

and

$$\lambda_1 \mu_i \geq 0, \qquad \lambda_2 \mu_i \geq 0, \qquad \sum_{i=1}^{n} \lambda_1 \mu_i + \sum_{i=n+1}^{m} \lambda_2 \mu_i = 1.$$

This shows that $\lambda_1 x_1 + \lambda_2 x_2$ is a convex combination of the points $y_1, \ldots, y_m$ from M, whence $\lambda_1 x_1 + \lambda_2 x_2$ is in C, as desired. □

Up to now we have had complete analogy with Section 1. The concept of a basis of a linear or affine subspace, however, has no analogue for convex sets in general. Still, we have the following substitute:

Theorem 2.3. *For any subset M of $\mathbb{R}^d$, the convex hull* conv M *is the set of all convex combinations*

$$\sum_{i=1}^{n}{}^{c} \lambda_i x_i$$

such that $(x_1, \ldots, x_n)$ is an affinely independent family of points from M.

In other words, in order to generate conv M we need not take *all* convex combinations of points from M as described by Theorem 2.2; it suffices to take those formed by the affinely independent families of points from M. On the other hand, no *fixed* family of points from M will suffice, as in the case of span M or aff M, cf. (d) and (d′) of Section 1.

Proof. We shall prove that if a point x is a convex combination of n points $x_1, \ldots, x_n$ such that $(x_1, \ldots, x_n)$ is affinely dependent, then x is already a convex combination of $n - 1$ of the points $x_1, \ldots, x_n$. Repeating this argument, if necessary, it follows that there is an affinely independent subfamily $(x_{i_1}, \ldots, x_{i_p})$ of $(x_1, \ldots, x_n)$ such that x is a convex combination of $x_{i_1}, \ldots, x_{i_p}$. The statement then follows from Theorem 2.2.

So, suppose that we have

$$x = \sum_{i=1}^{n}{}^{c} \lambda_i x_i, \tag{1}$$

where $(x_1, \ldots, x_n)$ is affinely dependent. If some λ_i is 0, then x is already a convex combination of $n - 1$ of the points $x_1, \ldots, x_n$. Hence, we may assume that all λ_i are >0. The affine dependence means that there are reals $\mu_1, \ldots, \mu_n$, not all 0, such that

$$\sum_{i=1}^{n} \mu_i x_i = o, \qquad \sum_{i=1}^{n} \mu_i = 0. \tag{2}$$

Combining (1) and (2) we see that for any real t we have

$$x = \sum_{i=1}^{n} (\lambda_i - t\mu_i)x_i \tag{3}$$

and

$$\sum_{i=1}^{n} (\lambda_i - t\mu_i) = 1.$$

We now simply seek a value of t (in fact, a positive value) such that $\lambda_i - t\mu_i \geq 0$ for all i, and $\lambda_i - t\mu_i = 0$ for at least one i; then (3) will be a representation of x as a convex combination of $n - 1$ of the points $x_1, \ldots, x_n$. We have $\lambda_i - t\mu_i > 0$ for any $t > 0$ when $\mu_i \leq 0$. When $\mu_i > 0$, we have $\lambda_i - t\mu_i \geq 0$ provided that $t \leq \lambda_i/\mu_i$, with $\lambda_i - t\mu_i = 0$ if and only if $t = \lambda_i/\mu_i$. Noting that we must have $\mu_i > 0$ for at least one i, we see that

$$t := \min\{\lambda_i/\mu_i \mid \mu_i > 0\}$$

fulfils the requirements. □

The following two corollaries are both known as *Carathéodory's Theorem*:

Corollary 2.4. *For any subset M of $\mathbb{R}^d$ with* dim(aff M) $= n$, *the convex hull* conv M *is the set of all convex combinations of at most $n + 1$ points from M.*

Proof. For any affinely independent m-family $(x_1, \ldots, x_m)$ of points from M, we have $m \leq n + 1$ by the assumption. Therefore, the set of all convex combinations of $n + 1$ or fewer points from M contains conv M by Theorem 2.3. On the other hand, it is contained in conv M by Theorem 2.2. □

Corollary 2.5. *For any subset M of $\mathbb{R}^d$ with* dim(aff M) $= n$, *the convex hull* conv M *is the set of all convex combinations of precisely $n + 1$ points from M.*

Proof. In a convex combination one may always add terms of the form $0x$. Therefore, the statement follows from Corollary 2.4. □

By a *convex polytope*, or simply a *polytope*, we mean a set which is the convex hull of a non-empty finite set $\{x_1, \ldots, x_n\}$. If P is a polytope, then any translate $x + P$ of P is also a polytope; this follows from the fact that $x +$ conv $M =$ conv$(x + M)$. More generally, the image of a polytope under an affine mapping is again a polytope; this follows from the fact that $\varphi(\text{conv } M) = \text{conv } \varphi(M)$ when φ is an affine mapping.

A polytope S with the property that there exists an affinely independent family $(x_1, \ldots, x_n)$ such that $S = \text{conv}\{x_1, \ldots, x_n\}$ is called a *simplex* (and the points $x_1, \ldots, x_n$ are called the *vertices* of S; cf. the remarks following Theorem 7.1).

One might say that simplices have a "convex basis," cf. the remark preceding Theorem 2.3. In fact, if $x_1, \ldots, x_n$ are the vertices of a simplex S, then by the affine independence each point in aff$\{x_1, \ldots, x_n\}$ has a unique representation as an affine combination of $x_1, \ldots, x_n$, whence, in particular, each point in $\operatorname{conv}\{x_1, \ldots, x_n\}$ has a unique representation as a convex combination of $x_1, \ldots, x_n$, cf. Theorem 2.3.

Convex sets having a "convex basis" in the sense described above must, of course, be polytopes. The following theorem shows that simplices are the only polytopes having a "convex basis":

Theorem 2.6. *Let $M = \{x_1, \ldots, x_n\}$ be a finite set of n points from $\mathbb{R}^d$ such that the n-family $(x_1, \ldots, x_n)$ is affinely dependent. Then there are subsets M_1 and M_2 of M with $M_1 \cap M_2 = \emptyset$ and $M_1 \cup M_2 = M$ such that*

$$\operatorname{conv} M_1 \cap \operatorname{conv} M_2 \neq \emptyset.$$

Proof. By the affine dependence there are reals $\lambda_1, \ldots, \lambda_n$, not all 0, such that

$$\sum_{i=1}^{n} \lambda_i x_i = o, \qquad \sum_{i=1}^{n} \lambda_i = 0. \tag{4}$$

Denoting the set $\{1, \ldots, n\}$ by I, we let

$$I_1 := \{i \in I \mid \lambda_i > 0\}, \qquad I_2 := \{i \in I \mid \lambda_i \leq 0\},$$

and we let

$$M_1 := \{x_i \mid i \in I_1\}, \qquad M_2 := \{x_i \mid i \in I_2\}.$$

Now, take

$$x := \sum_{i \in I_1} \frac{\lambda_i}{\lambda} x_i, \tag{5}$$

where

$$\lambda := \sum_{i \in I_1} \lambda_i.$$

(It is clear that $I_1 \neq \emptyset$, whence $\lambda > 0$.) The right-hand side of (5) is in fact a convex combination, whence x is in $\operatorname{conv} M_1$ by Theorem 2.2. However, using (4) we see that we also have

$$x = \sum_{i \in I_2} \frac{-\lambda_i}{\lambda} x_i,$$

and again we actually have a convex combination. Therefore, x is also in $\operatorname{conv} M_2$. Consequently, $\operatorname{conv} M_1$ and $\operatorname{conv} M_2$ have the point x in common. □

The following corollary of Theorem 2.6 is known as *Radon's Theorem*:

Corollary 2.7. *Let $M = \{x_1, \ldots, x_n\}$ be a finite set of n points from $\mathbb{R}^d$ such that $n \geq d + 2$. Then there are subsets M_1 and M_2 of M with $M_1 \cap M_2 = \emptyset$ and $M_1 \cup M_2 = M$ such that*

$$\operatorname{conv} M_1 \cap \operatorname{conv} M_2 \neq \emptyset.$$

PROOF. The maximum number of members in an affinely independent family of points from $\mathbb{R}^d$ is $d + 1$. Therefore, $(x_1, \ldots, x_n)$ must be affinely dependent, whence Theorem 2.6 applies. □

We conclude this section with an important application of Carathéodory's Theorem.

Theorem 2.8. *For any compact subset M of $\mathbb{R}^d$, the convex hull* conv M *is again compact.*

PROOF. Let $(y_\nu)_{\nu \in \mathbb{N}}$ be any sequence of points from conv M. We shall prove that the sequence admits a subsequence which converges to a point in conv M. Let the dimension of aff M be denoted by n. Then Corollary 2.5 shows that each y_ν in the sequence has a representation

$$y_\nu = \sum_{i=1}^{n+1}{}^{c} \lambda_{\nu i} x_{\nu i},$$

where $x_{\nu i} \in M$. We now consider the $n + 1$ sequences

$$(x_{\nu 1})_{\nu \in \mathbb{N}}, \ldots, (x_{\nu(n+1)})_{\nu \in \mathbb{N}} \tag{6}$$

of points from M, and the $n + 1$ sequences

$$(\lambda_{\nu 1})_{\nu \in \mathbb{N}}, \ldots, (\lambda_{\nu(n+1)})_{\nu \in \mathbb{N}} \tag{7}$$

of real numbers from $[0, 1]$. By the compactness of M there is a subsequence of $(x_{\nu 1})_{\nu \in \mathbb{N}}$ which converges to a point in M. Replace all $2(n + 1)$ sequences by the corresponding subsequences. Change notation such that (6) and (7) now denote the subsequences; then $(x_{\nu 1})_{\nu \in \mathbb{N}}$ converges in M. Next, use the compactness of M again to see that there is a subsequence of the (sub)sequence $(x_{\nu 2})_{\nu \in \mathbb{N}}$ which converges to a point in M. Change notation, etc. Then after $2(n + 1)$ steps, where we use the compactness of M in step $1, \ldots, n + 1$, and the compactness of $[0, 1]$ in step $n + 2, \ldots, 2n + 2$, we end up with subsequences

$$(x_{\nu_m 1})_{m \in \mathbb{N}}, \ldots, (x_{\nu_m(n+1)})_{m \in \mathbb{N}}$$

of the original sequences (6) which converge in M, say

$$\lim_{m \to \infty} x_{\nu_m i} = x_{0i}, \qquad i = 1, \ldots, n + 1,$$

and subsequences

$$(\lambda_{\nu_m 1})_{m \in \mathbb{N}}, \ldots, (\lambda_{\nu_m(n+1)})_{m \in \mathbb{N}}$$

of the original sequences (7) which converge in $[0, 1]$, say

$$\lim_{m \to \infty} \lambda_{\nu_m i} = \lambda_{0i}, \qquad i = 1, \ldots, n + 1.$$

Since

$$\sum_{i=1}^{n+1} \lambda_{\nu_m i} = 1, \qquad m \in \mathbb{N},$$

we also have

$$\sum_{i=1}^{n+1} \lambda_{0i} = 1.$$

Then the linear combination

$$y_0 := \sum_{i=1}^{n+1} \lambda_{0i} x_{0i}$$

is in fact a convex combination. Therefore, y_0 is in conv M by Theorem 2.2. It is also clear that

$$\lim_{m \to \infty} y_{\nu_m} = y_0 .$$

In conclusion, $(y_{\nu_m})_{m \in \mathbb{N}}$ is a subsequence of $(y_\nu)_{\nu \in \mathbb{N}}$ which converges to a point in conv M. □

Some readers may prefer the following version of the proof above. With $n = \dim(\operatorname{aff} M)$ as above, let

$$S := \{(\lambda_1, \ldots, \lambda_{n+1}) \in \mathbb{R}^{n+1} \mid \lambda_1, \ldots, \lambda_{n+1} \geq 0, \lambda_1 + \cdots + \lambda_{n+1} = 1\},$$

and define a mapping $\varphi: M^{n+1} \times S \to \mathbb{R}^d$ by

$$\varphi((x_1, \ldots, x_{n+1}), (\lambda_1, \ldots, \lambda_{n+1})) := \sum_{i=1}^{n+1} \lambda_i x_i .$$

By Corollary 2.5, the set $\varphi(M^{n+1} \times S)$ is precisely conv M. Now, $M^{n+1} \times S$ is compact by the compactness of M and S, and φ is continuous. Since the continuous image of a compact set is again compact, it follows that conv M is compact.

Since any finite set is compact, Theorem 2.8 immediately implies:

Corollary 2.9. *Any convex polytope P in $\mathbb{R}^d$ is a compact set.*

One should observe, however, that a direct proof of Corollary 2.9 does not require Carathéodory's Theorem. In fact, if M is the finite set $\{x_1, \ldots, x_m\}$, then each y_ν (in the notation of the proof above) has a representation

$$y_\nu = \sum_{i=1}^{m}{}^{c} \lambda_{\nu i} x_i .$$

Then we have a similar situation as in the proof above (with m corresponding to $n + 1$), except that now the sequences corresponding to the sequences (6) are constant, $x_{\nu i} = x_i$ for all ν. Therefore, we need only show here that the sequences (7) admit converging subsequences (which is proved as above).

Exercises

2.1. Show that when C_1 and C_2 are convex sets in $\mathbb{R}^d$, then the set

$$C_1 + C_2 := \{x_1 + x_2 \mid x_1 \in C_1, x_2 \in C_2\}$$

is also convex.

2.2. Show that when C is a convex set in $\mathbb{R}^d$, and λ is a real, then the set

$$\lambda C := \{\lambda x \mid x \in C\}$$

is also convex.

2.3. Show that when C is a convex set in $\mathbb{R}^d$, and $\varphi: \mathbb{R}^d \to \mathbb{R}^e$ is an affine mapping, then $\varphi(C)$ is also convex.

2.4. Show that $\operatorname{conv}(M_1 + M_2) = \operatorname{conv} M_1 + \operatorname{conv} M_2$ for any subsets M_1 and M_2 of $\mathbb{R}^d$.

2.5. Show that when M is any subset of $\mathbb{R}^d$, and $\varphi: \mathbb{R}^d \to \mathbb{R}^e$ is an affine mapping, then $\varphi(\operatorname{conv} M) = \operatorname{conv} \varphi(M)$. Deduce in particular that the affine image of a polytope is again a polytope.

2.6. Show that when M is an open subset of $\mathbb{R}^d$, then conv M is also open. Use this fact to show that the interior of a convex set is again convex. (Cf. Theorem 3.4(b).)

2.7. Show by an example in $\mathbb{R}^2$ that the convex hull of a closed set need not be closed. (Cf. Theorem 2.8.)

2.8. An n-family $(x_1, \ldots, x_n)$ of points from $\mathbb{R}^d$ is said to be *convexly independent* if no x_i in the family is a convex combination of the remaining x_j's. For $n \geq d + 2$, show that if every $(d + 2)$-subfamily of $(x_1, \ldots, x_n)$ is convexly independent, then the entire n-family is convexly independent.

2.9. Let $(C_i)_{i \in I}$ be a family of convex sets in $\mathbb{R}^d$ with $d + 1 \leq \operatorname{card} I$. Consider the following two statements:

(a) Any $d + 1$ of the sets C_i have a non-empty intersection.
(b) All the sets C_i have a non-empty intersection.

Prove *Helly's Theorem*: If $\operatorname{card} I < \infty$, then (a) $\Rightarrow$ (b). (Hint: Use induction on $n := \operatorname{card} I$. Apply Corollary 2.7.)

Show by an example that we need not have (a) $\Rightarrow$ (b) when $\operatorname{card} I = \infty$.

Prove that if each C_i is closed, and at least one is compact, then we have (a) $\Rightarrow$ (b) without restriction on card I.

2.10. Let a point x in $\mathbb{R}^d$ be a convex combination of points $x_1, \ldots, x_n$, and let each x_i be a convex combination of points $y_{i1}, \ldots, y_{in_i}$. Show that x is a convex combination of the points $y_{i\nu_i}$, $i = 1, \ldots, n$, $\nu_i = 1, \ldots, n_i$.

2.11. Let $(C_i)_{i \in I}$ be a family of distinct convex sets in $\mathbb{R}^d$. Show that

$$\operatorname{conv} \bigcup_{i \in I} C_i$$

is the set of all convex combinations

$$\sum_{\nu=1}^{n}{}^{c} \lambda_{i_\nu} x_{i_\nu},$$

where $x_{i_\nu} \in C_{i_\nu}$.

Deduce in particular that when C_1 and C_2 are convex, then $\operatorname{conv}(C_1 \cup C_2)$ is the union of all segments $[x_1, x_2]$ with $x_1 \in C_1$ and $x_2 \in C_2$.

§3. The Relative Interior of a Convex Set

It is clear that the interior of a convex set may be empty. A triangle in $\mathbb{R}^3$, for example, has no interior points. However, it does have interior points in the 2-dimensional affine space that it spans. This observation illustrates the definition below of the relative interior of a convex set, and the main result of this section, Theorem 3.1. We shall also discuss the behaviour of a convex set under the operations of forming (relative) interior, closure, and boundary.

By the *relative interior* of a convex set C in $\mathbb{R}^d$ we mean the interior of C in the affine hull aff C of C. The relative interior of C is denoted by ri C. Points in ri C are called *relative interior points* of C. The set cl $C \setminus$ ri C is called the *relative boundary* of C, and is denoted by rb C. Points in rb C are called *relative boundary points* of C. (Since aff C is a closed subset of $\mathbb{R}^d$, the "relative closure" of C is simply the closure of C. Hence, the relative boundary of C is actually the boundary of C in aff C.)

It should be noted that the ri-operation is not just a slight modification of the int-operation. Most striking, perhaps, is the fact that the ri-operation does not preserve inclusions. For example, let C_1 be a side of a triangle C_2. Then $C_1 \subset C_2$, whereas ri $C_1 \not\subset$ ri C_2; in fact, ri C_1 and ri C_2 are non-empty disjoint sets.

By the *dimension* of a convex set C we mean the dimension dim(aff C) of the affine hull of C; it is denoted by dim C. The empty set has dimension -1. The 0-dimensional convex sets are the 1-point sets $\{x\}$. The 1-dimensional convex sets are the (closed, half-open or open) segments, the (closed or open) halflines, and the lines.

For a 0-dimensional convex set $C = \{x\}$, we clearly have ri $C = C$, cl $C = C$, and rb $C = \emptyset$.

We have ri C = int C for a non-empty convex set C in $\mathbb{R}^d$ if and only if int $C \neq \emptyset$. In fact, if int $C \neq \emptyset$ then aff $C = \mathbb{R}^d$, whence ri C = int C by the definition of ri C. The converse is a consequence of the following:

Theorem 3.1. *Let C be any non-empty convex set in $\mathbb{R}^d$. Then* ri $C \neq \emptyset$.

We first prove Theorem 3.1 for simplices:

Lemma 3.2. *Let S be a simplex in $\mathbb{R}^d$. Then* ri $S \neq \emptyset$.

PROOF. When dim $S = k$, there is a $(k + 1)$-family $(x_1, \ldots, x_{k+1})$, affinely independent, such that

$$S = \operatorname{conv}\{x_1, \ldots, x_{k+1}\}.$$

Then $(x_1, \ldots, x_{k+1})$ is an affine basis of aff S; hence, aff S is the set of points of the form

$$x = \sum_{i=1}^{k+1}{}^{a} \lambda_i x_i,$$

and for each $x \in \operatorname{aff} S$, the coefficients $\lambda_1, \ldots, \lambda_{k+1}$ are unique. Therefore, we may define a mapping

$$\varphi: \operatorname{aff} S \to \mathbb{R}^{k+1}$$

by letting

$$\varphi\left(\sum_{i=1}^{k+1}{}^{a} \lambda_i x_i\right) := (\lambda_1, \ldots, \lambda_{k+1}).$$

This is actually an affine mapping; in particular, it is continuous. Let

$$K_i := \{(\lambda_1, \ldots, \lambda_{k+1}) \in \mathbb{R}^{k+1} \mid \lambda_i > 0\}, \qquad i = 1, \ldots, k+1.$$

Then $K_1, \ldots, K_{k+1}$ are open halfspaces in $\mathbb{R}^{k+1}$, and therefore, by continuity, the sets $\varphi^{-1}(K_1), \ldots, \varphi^{-1}(K_{k+1})$ are open (in fact, open halfspaces) in $\operatorname{aff} S$. The set

$$\bigcap_{i=1}^{k+1} \varphi^{-1}(K_i), \tag{1}$$

is therefore also open in $\operatorname{aff} S$. Now, note that

$$\bigcap_{i=1}^{k+1} \varphi^{-1}(K_i) = \left\{\sum_{i=1}^{k+1}{}^{a} \lambda_i x_i \,\middle|\, \lambda_1, \ldots, \lambda_{k+1} > 0\right\}.$$

This shows in particular that the set (1) is non-empty. And since affine combinations $\lambda_1 x_1 + \cdots + \lambda_{k+1} x_{k+1}$ with all $\lambda_i > 0$ are convex combinations, we see that the set (1) is a subset of S. In other words, the set S contains a non-empty set which is open in $\operatorname{aff} S$, whence $\operatorname{ri} S \neq \varnothing$. (The proof shows that the set (1) is a subset of $\operatorname{ri} S$. Actually, the two sets are the same.) □

With Lemma 3.2 at hand we can now pass to:

Proof (Theorem 3.1). Let

$$k := \dim C \; (= \dim(\operatorname{aff} C)).$$

Then there is an affinely independent $(k+1)$-family $(x_1, \ldots, x_{k+1})$ of points from C (but no such $(k+2)$-family). Let

$$S := \operatorname{conv}\{x_1, \ldots, x_{k+1}\}.$$

Then S is a simplex contained in C. By Lemma 3.2, S has a non-empty interior relative to $\operatorname{aff} S$. Since

$$\operatorname{aff} S \subset \operatorname{aff} C$$

and

$$\dim(\operatorname{aff} S) = k = \dim(\operatorname{aff} C),$$

we actually have

$$\operatorname{aff} S = \operatorname{aff} C.$$

Therefore, S has a non-empty interior relative to aff C. But since S is a subset of C, it follows that C has a non-empty interior relative to aff C, as desired. □

The following theorem shows that any point in the closure of a convex set C can be "seen" from any relative interior point of C "via" relative interior points:

Theorem 3.3. *Let C be a convex set in $\mathbb{R}^d$. Then for any $x_0 \in \operatorname{ri} C$ and any $x_1 \in \operatorname{cl} C$ with $x_0 \neq x_1$ we have $[x_0, x_1[\subset \operatorname{ri} C$.*

PROOF. It is easy to prove the statement in the particular case where we have $x_0 \in \operatorname{int} C$ and $x_1 \in C$. For $\lambda \in]0, 1[$, let $x_\lambda := (1 - \lambda)x_0 + \lambda x_1$. From $x_0 \in \operatorname{int} C$ it follows that there is a ball B centred at x_0 with $B \subset C$. From $x_1 \in C$ and the convexity of C it next follows that

$$B_\lambda := (1 - \lambda)B + \lambda x_1$$

is contained in C. But B_λ is a ball centred at x_λ, whence $x_\lambda \in \operatorname{int} C$, as desired. The proof below covering the general case is an elaborate version of this idea. Of course, the main difficulty is that x_1 need not be in C.

So, consider $x_0 \in \operatorname{ri} C$ and $x_1 \in \operatorname{cl} C$ with $x_0 \neq x_1$. For any $\lambda \in]0, 1[$, let

$$x_\lambda := (1 - \lambda)x_0 + \lambda x_1.$$

We shall prove that $x_\lambda \in \operatorname{ri} C$. Since x_0 is a relative interior point of C, there is a (relatively) open subset U of aff C such that $x_0 \in U \subset C$. Let

$$V := \lambda^{-1}(x_\lambda - (1 - \lambda)U).$$

Since

$$\lambda^{-1} - \lambda^{-1}(1 - \lambda) = 1,$$

it follows that V is a subset of aff C, and it is, in fact, (relatively) open. And since

$$x_1 = \lambda^{-1}(x_\lambda - (1 - \lambda)x_0),$$

we see that $x_1 \in V$. Therefore, by the assumption that $x_1 \in \operatorname{cl} C$, there is a point $y_1 \in V \cap C$. Let

$$W := (1 - \lambda)U + \lambda y_1.$$

Then W is a (relatively) open subset of aff C, and since we have both $U \subset C$ and $y_1 \in C$, it follows that $W \subset C$ by the convexity of C. We complete the proof by showing that $x_\lambda \in W$. From the definition of V it follows that there is a point $y_0 \in U$ such that

$$y_1 = \lambda^{-1}(x_\lambda - (1 - \lambda)y_0).$$

Then

$$\begin{aligned} x_\lambda &= (1 - \lambda)y_0 + \lambda y_1 \\ &\in (1 - \lambda)U + \lambda y_1 = W, \end{aligned}$$

as desired. □

Theorem 3.3 is a useful tool. Among other things, it is crucial for the proofs of all the statements, except (a), in the following theorem. The theorem brings out the nice behaviour of convex sets.

Theorem 3.4. *For any convex set C in $\mathbb{R}^d$ one has:*

(a) $\operatorname{cl} C$ *is convex.*
(b) $\operatorname{ri} C$ *is convex.*
(c) $\operatorname{cl} C = \operatorname{cl}(\operatorname{cl} C) = \operatorname{cl}(\operatorname{ri} C)$.
(d) $\operatorname{ri} C = \operatorname{ri}(\operatorname{cl} C) = \operatorname{ri}(\operatorname{ri} C)$.
(e) $\operatorname{rb} C = \operatorname{rb}(\operatorname{cl} C) = \operatorname{rb}(\operatorname{ri} C)$.
(f) $\operatorname{aff} C = \operatorname{aff}(\operatorname{cl} C) = \operatorname{aff}(\operatorname{ri} C)$.
(g) $\dim C = \dim(\operatorname{cl} C) = \dim(\operatorname{ri} C)$.

PROOF. For $C = \emptyset$, there is nothing to prove. So, we may assume that C is non-empty, whenever necessary.

(a) Let $x_0, x_1 \in \operatorname{cl} C$, and let $\lambda \in]0, 1[$. We shall prove that the point

$$x_\lambda := (1 - \lambda)x_0 + \lambda x_1$$

is also in $\operatorname{cl} C$. Now, there are sequences

$$(x_{0\nu})_{\nu \in \mathbb{N}}, \qquad (x_{1\nu})_{\nu \in \mathbb{N}}$$

of points from C such that

$$\lim_{\nu \to \infty} x_{0\nu} = x_0, \qquad \lim_{\nu \to \infty} x_{1\nu} = x_1.$$

By the convexity of C, the points

$$(1 - \lambda)x_{0\nu} + \lambda x_{1\nu}, \qquad \nu \in \mathbb{N},$$

are all in C. Furthermore,

$$\lim_{\nu \to \infty} ((1 - \lambda)x_{0\nu} + \lambda x_{1\nu}) = (1 - \lambda)x_0 + \lambda x_1 = x_\lambda.$$

This shows that $x_\lambda \in \operatorname{cl} C$.

(b) We shall prove that for any $x_0, x_1 \in \operatorname{ri} C$ and any $\lambda \in]0, 1[$, the point

$$x_\lambda := (1 - \lambda)x_0 + \lambda x_1$$

is also in $\operatorname{ri} C$. This follows immediately from Theorem 3.3.

(c) The statement $\operatorname{cl} C = \operatorname{cl}(\operatorname{cl} C)$ is trivial. It is also trivial that $\operatorname{cl}(\operatorname{ri} C) \subset \operatorname{cl} C$. To prove the opposite inclusion, let $x_1 \in \operatorname{cl} C$. Take any point $x_0 \in \operatorname{ri} C$, cf. Theorem 3.1. If $x_0 = x_1$, then we have

$$x_1 \in \operatorname{ri} C \subset \operatorname{cl}(\operatorname{ri} C),$$

as desired. If $x_0 \neq x_1$, then we have

$$[x_0, x_1[\subset \operatorname{ri} C,$$

cf. Theorem 3.3. Since each neighbourhood of x_1 contains points from $[x_0, x_1[$, it follows that x_1 is in cl(ri C).

(d) To prove that ri C = ri(cl C), we first note that

$$\operatorname{aff} C = \operatorname{aff}(\operatorname{cl} C), \tag{2}$$

since aff C is closed. Then it is clear that ri $C \subset$ ri(cl C). To prove the opposite inclusion, let x be in ri(cl C). Take any point $x_0 \in \operatorname{ri} C$, cf. Theorem 3.1. If $x_0 = x$, then we have $x \in \operatorname{ri} C$, as desired. If $x_0 \neq x$, then $\operatorname{aff}\{x_0, x\}$ is a line, and we have

$$\operatorname{aff}\{x_0, x\} \subset \operatorname{aff}(\operatorname{cl} C) = \operatorname{aff} C.$$

Since $x \in \operatorname{ri}(\operatorname{cl} C)$, there is a point $x_1 \in \operatorname{aff}\{x_0, x\}$ such that $x_1 \in \operatorname{cl} C$ and $x \in]x_0, x_1[$. Application of Theorem 3.3 then yields $x \in \operatorname{ri} C$. Hence, ri C = ri(cl C).

To prove that ri C = ri(ri C), we first verify that

$$\operatorname{aff} C = \operatorname{aff}(\operatorname{ri} C). \tag{3}$$

Applying (2) to ri C instead of C and using (c), we obtain

$$\begin{aligned} \operatorname{aff}(\operatorname{ri} C) &= \operatorname{aff}(\operatorname{cl}(\operatorname{ri} C)) \\ &= \operatorname{aff}(\operatorname{cl} C) \\ &= \operatorname{aff} C. \end{aligned}$$

Now, using the notation $\operatorname{int}_{\operatorname{aff} C} C$ for ri C, we have

$$\begin{aligned} \operatorname{ri}(\operatorname{ri} C) &= \operatorname{int}_{\operatorname{aff}(\operatorname{ri} C)}(\operatorname{ri} C) \\ &= \operatorname{int}_{\operatorname{aff} C}(\operatorname{ri} C), \end{aligned}$$

where we have used (3). But

$$\begin{aligned} \operatorname{int}_{\operatorname{aff} C}(\operatorname{ri} C) &= \operatorname{int}_{\operatorname{aff} C}(\operatorname{int}_{\operatorname{aff} C} C) \\ &= \operatorname{int}_{\operatorname{aff} C} C \\ &= \operatorname{ri} C, \end{aligned}$$

where we have used the standard fact that int(int M) = int M for any set M. This completes the proof of (d).

(e) By definition we have

$$\operatorname{rb} C = \operatorname{cl} C \setminus \operatorname{ri} C,$$

$$\operatorname{rb}(\operatorname{cl} C) = \operatorname{cl}(\operatorname{cl} C) \setminus \operatorname{ri}(\operatorname{cl} C),$$

$$\operatorname{rb}(\operatorname{ri} C) = \operatorname{cl}(\operatorname{ri} C) \setminus \operatorname{ri}(\operatorname{ri} C).$$

The statement then follows using (c) and (d).

(f) This has already been proved, cf. (2) and (3) above.

(g) This follows from (f). □

The next theorem also depends on Theorem 3.3. It shows that the relative interior points of a convex set C may be characterized in purely algebraic terms:

Theorem 3.5. *For any convex set C in $\mathbb{R}^d$ and any point $x \in C$ the following three conditions are equivalent:*

(a) $x \in \operatorname{ri} C$.
(b) *For any line A in* aff C *with $x \in A$ there are points $y_0, y_1 \in A \cap C$ such that $x \in]y_0, y_1[$.*
(c) *For any point $y \in C$ with $y \neq x$ there is a point $z \in C$ such that $x \in]y, z[$, i.e. any segment $[y, x]$ in C can be extended beyond x in C.*

PROOF. The implications (a) $\Rightarrow$ (b) and (b) $\Rightarrow$ (c) are obvious. Therefore, we need only prove (c) $\Rightarrow$ (a). By Theorem 3.1 there is a point $y \in \operatorname{ri} C$. If $y = x$, there is nothing more to prove. If $y \neq x$, then by (c) there is a point $z \in C$ such that $x \in]y, z[$. But then x is in ri C by Theorem 3.3. □

We conclude this section with an application of Theorem 3.4(a). Let M be any set in $\mathbb{R}^d$. Then there is a smallest closed convex set containing M, namely, the intersection of all closed convex sets containing M. We call this set the *closed convex hull* of M, and denote it by clconv M. As might be expected, we have:

Theorem 3.6. *Let M be any subset of $\mathbb{R}^d$. Then*

$$\operatorname{clconv} M = \operatorname{cl}(\operatorname{conv} M),$$

i.e. the closed convex hull of M is the closure of the convex hull of M.

PROOF. Using Theorem 3.4(a) we see that cl(conv M) is a closed convex set containing M. Since clconv M is the smallest such set, it follows that

$$\operatorname{clconv} M \subset \operatorname{cl}(\operatorname{conv} M).$$

On the other hand, clconv M is a convex set containing M, whence

$$\operatorname{clconv} M \supset \operatorname{conv} M.$$

Since clconv M is also closed, this implies

$$\operatorname{clconv} M \supset \operatorname{cl}(\operatorname{conv} M),$$

completing the proof. □

EXERCISES

3.1. Let $P = \operatorname{conv}\{x_1, \ldots, x_n\}$ be a polytope in $\mathbb{R}^d$. Show that a point x is in ri P if and only if x is a convex combination of $x_1, \ldots, x_n$ with strictly positive coefficients, i.e. there are $\lambda_1, \ldots, \lambda_n$ such that

$$x = \sum_{i=1}^{n}{}^{c} \lambda_i x_i$$

and $\lambda_i > 0$ for $i = 1, \ldots, n$.

3.2. Let C_1 and C_2 be convex sets in $\mathbb{R}^d$. Show that $\operatorname{ri}(C_1 + C_2) = \operatorname{ri} C_1 + \operatorname{ri} C_2$.

3.3. Let C be a convex set in $\mathbb{R}^d$, and let $\varphi: \mathbb{R}^d \to \mathbb{R}^e$ be an affine mapping. Show that $\operatorname{ri} \varphi(C) = \varphi(\operatorname{ri} C)$.

3.4. Let $(C_i)_{i \in I}$ be a family of convex sets in $\mathbb{R}^d$ such that

$$\bigcap_{i \in I} \operatorname{ri} C_i \neq \varnothing. \tag{4}$$

Show that

$$\operatorname{cl} \bigcap_{i \in I} C_i = \bigcap_{i \in I} \operatorname{cl} C_i. \tag{5}$$

Show that if (4) does not hold, then (5) need not hold.

3.5. Let $(C_i)_{i=1,\ldots,n}$ be a finite family of convex sets in $\mathbb{R}^d$ such that

$$\bigcap_{i=1}^{n} \operatorname{ri} C_i \neq \varnothing. \tag{6}$$

Show that

$$\operatorname{ri} \bigcap_{i=1}^{n} C_i = \bigcap_{i=1}^{n} \operatorname{ri} C_i. \tag{7}$$

Show that if (6) does not hold, then (7) need not hold.

§4. Supporting Hyperplanes and Halfspaces

It is intuitively clear that when x is a relative boundary point of a convex set C, then there is a hyperplane H passing through x such that all points of C not in H are on the same side of H. One of the main results of this section shows that it is in fact so.

Let C be a non-empty closed convex set in $\mathbb{R}^d$. By a *supporting halfspace* of C we mean a closed halfspace K in $\mathbb{R}^d$ such that $C \subset K$ and $H \cap C \neq \varnothing$, where H denotes the bounding hyperplane of K. By a *supporting hyperplane* of C we mean a hyperplane H in $\mathbb{R}^d$ which bounds a supporting halfspace.

In the definition of a supporting hyperplane H of C we allow $C \subset H$ (in which case both closed halfspaces bounded by H are supporting halfspaces). If C is not contained in H we shall call H a *proper supporting hyperplane*.

Analytically, a hyperplane $H(y, \alpha)$ is a supporting hyperplane of a non-empty closed convex set C if and only if

$$\alpha = \max_{x \in C} \langle x, y \rangle \tag{1}$$

or

$$\alpha = \min_{x \in C} \langle x, y \rangle. \tag{2}$$

If (2) holds for $H(y, \alpha)$, then (1) holds for $H(-y, -\alpha)$. Since $H(-y, -\alpha) = H(y, \alpha)$, it follows that any supporting hyperplane H of C has the form $H(y, \alpha)$ such that (1) holds, whence $C \subset K(y, \alpha)$. Note, by the way, that if $H(y, \alpha)$ is a supporting hyperplane such that $C \subset K(y, \alpha)$, then $H(y, \alpha)$ is proper if and only if

$$\inf_{x \in C} \langle x, y \rangle < \max_{x \in C} \langle x, y \rangle.$$

We first prove:

Theorem 4.1. *Let C be a non-empty convex set in $\mathbb{R}^d$, and let H be a hyperplane in $\mathbb{R}^d$. Then the following two conditions are equivalent*:

(a) $H \cap \operatorname{ri} C = \emptyset$.
(b) *C is contained in one of the two closed halfspaces bounded by H, but not in H.*

Proof. Assume that (a) holds. Let $x_0 \in \operatorname{ri} C$, cf. Theorem 3.1. Then $x_0 \notin H$ by (a). In particular, C is not contained in H. Suppose that there is a point $x_1 \in C$ such that x_0 and x_1 are on opposite sides of H. Then there are y and α such that $H = H(y, \alpha)$ and

$$\langle x_0, y \rangle < \alpha < \langle x_1, y \rangle.$$

Taking

$$\lambda := \frac{\alpha - \langle x_0, y \rangle}{\langle x_1, y \rangle - \langle x_0, y \rangle}$$

and

$$x_\lambda := (1 - \lambda)x_0 + \lambda x_1,$$

we have $\lambda \in]0, 1[$, and so $x_\lambda \in]x_0, x_1[$. Furthermore, an easy computation shows that $\langle x_\lambda, y \rangle = \alpha$, whence $x_\lambda \in H$. On the other hand, since $x_0 \in \operatorname{ri} C$ and $x_\lambda \in]x_0, x_1[$, it follows from Theorem 3.3 that we also have $x_\lambda \in \operatorname{ri} C$, whence $x_\lambda \in H \cap \operatorname{ri} C$, a contradiction. In conclusion, C is contained in that closed halfspace bounded by H which contains the point x_0.

Conversely, assume that (b) holds. Suppose that there is a point $x \in H \cap \operatorname{ri} C$. By (b) there is a point $y \in C \setminus H$. Then by Theorem 3.5, (a) $\Rightarrow$ (c) there is a point $z \in C$ such that $x \in]y, z[$, whence

$$x = (1 - \lambda)y + \lambda z$$

for a suitable $\lambda \in]0, 1[$. Now, there are u and α such that $H = H(u, \alpha)$ and $C \subset K(u, \alpha)$. Then $\langle y, u \rangle < \alpha$ and $\langle z, u \rangle \leq \alpha$, whence

$$\begin{aligned}\langle x, u \rangle &= \langle (1 - \lambda)y + \lambda z, u \rangle \\ &= (1 - \lambda)\langle y, u \rangle + \lambda \langle z, u \rangle \\ &< (1 - \lambda)\alpha + \lambda\alpha = \alpha.\end{aligned}$$

At the same time we have $\langle x, u \rangle = \alpha$ since $x \in H$, a contradiction. Therefore, $H \cap \operatorname{ri} C$ is empty. □

We immediately get:

Corollary 4.2. *A supporting hyperplane H of a non-empty closed convex set C in $\mathbb{R}^d$ is a proper supporting hyperplane of C if and only if $H \cap \operatorname{ri} C = \emptyset$.*

The following result is fundamental:

Theorem 4.3. *Let C be a closed convex set in $\mathbb{R}^d$, and let x be a point in* $\operatorname{rb} C$. *Then there is a proper supporting hyperplane H of C such that $x \in H$.*

We shall build the proof of Theorem 4.3 upon the following:

Lemma 4.4. *Let C be a non-empty open convex set in $\mathbb{R}^d$, and let x be a point of $\mathbb{R}^d$ not in C. Then there is a hyperplane H in $\mathbb{R}^d$ such that $x \in H$ and $H \cap C = \emptyset$.*

PROOF. We shall use induction on d. The statement is trivially true for $d = 0, 1$. We also need a proof for $d = 2$, however. So, let C be a non-empty open convex set in $\mathbb{R}^2$, and let $x \in \mathbb{R}^2 \setminus C$. We shall prove that there exists a line L in $\mathbb{R}^2$ such that $x \in L$ and $L \cap C = \emptyset$. Let S be a circle with its centre at x, and for each point $u \in C$ let u' be the unique point of S where the halfline

$$\{(1 - \lambda)x + \lambda u \mid \lambda > 0\}$$

from x through u meets S. Then the set

$$C' := \{u' \mid u \in C\}$$

is an open arc in S. Since $x \notin C$ and C is convex, two opposite points of S cannot both be in C'. Therefore, the angle between the two halflines from x through the endpoints of C' is at most π. Any of the two lines determined by one of these halflines can then be used as L. (If the angle is π, then, of course, L is unique.)

Next, let $d > 2$, and assume the statement is valid for all dimensions less than d. Let C be a non-empty open convex set in $\mathbb{R}^d$, and let $x \in \mathbb{R}^d \setminus C$. (See Figure 1 which illustrates the "difficult" situation where $x \in \operatorname{cl} C$.) Take any 2-dimensional affine subspace A of $\mathbb{R}^d$ such that $x \in A$ and $A \cap C \neq \emptyset$. Then $A \cap C$ is a non-empty open convex set in A with $x \notin A \cap C$. Identifying A with $\mathbb{R}^2$ and using the result on $\mathbb{R}^2$ proved above, we see that there exists a line L in A such that $x \in L$ and

$$L \cap (A \cap C) = L \cap C = \emptyset.$$

Let B be any hyperplane in $\mathbb{R}^d$ orthogonal to L, and let $\pi \colon \mathbb{R}^d \to B$ denote the orthogonal projection. Then $\pi(C)$ is a non-empty open convex set in B. Moreover, since $\pi^{-1}(\pi(x)) = L$, we see that $\pi(x) \notin \pi(C)$. Then, by hypothesis, there is a hyperplane H' in B such that $\pi(x) \in H'$ and $H' \cap \pi(C) = \emptyset$. But then

$$H := \operatorname{aff}(H' \cup L) = \pi^{-1}(H')$$

is a hyperplane in $\mathbb{R}^d$ with $x \in H$ and $H \cap C = \emptyset$. □

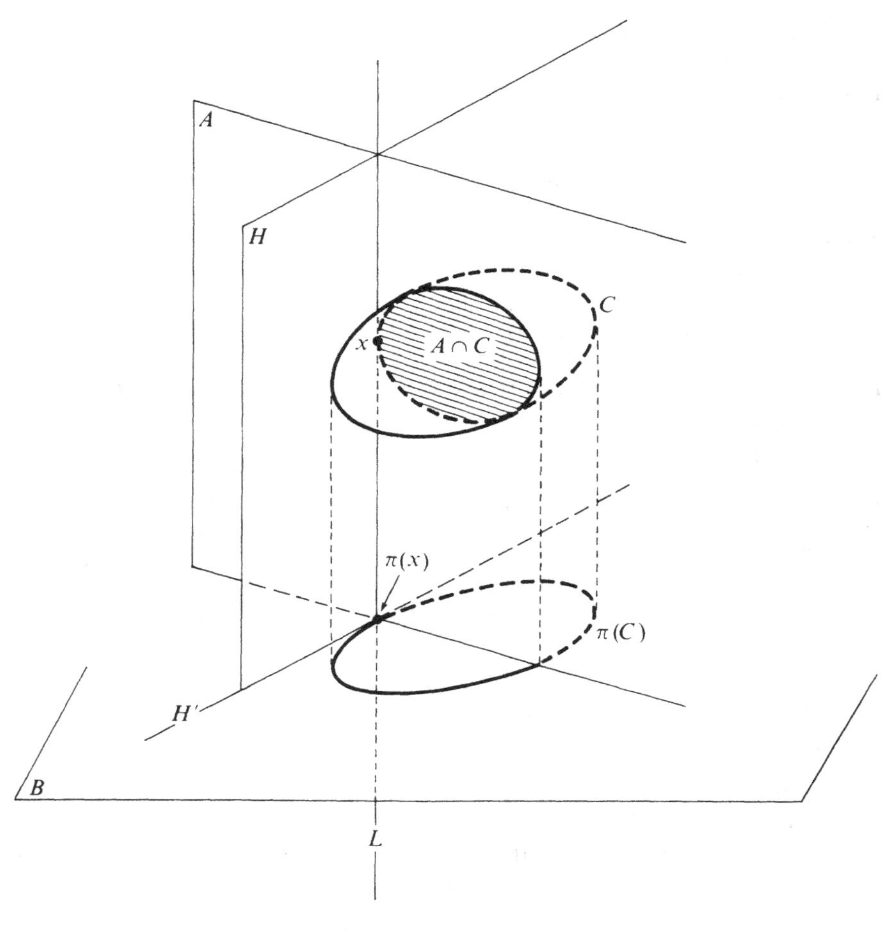

Figure 1

We can now prove Theorem 4.3:

PROOF (Theorem 4.3). When dim $C = -1, 0$, there is nothing to prove. So, let dim $C \geq 1$, and let C and x be as described. We shall apply Lemma 4.4 to the convex set ri C and the point x in the affine space aff C. (Here we need to identify aff C with $\mathbb{R}^e$ where $e := \dim(\operatorname{aff} C)$.) To see that Lemma 4.4 applies, note that ri C is non-empty by Theorem 3.1, convex by Theorem 3.4(b), and open in aff C; furthermore, x is in aff C. Application of Lemma 4.4 then yields the existence of a hyperplane H' in aff C such that $x \in H'$ and $H' \cap \operatorname{ri} C = \emptyset$. Clearly there is a hyperplane H in $\mathbb{R}^d$ such that $H \cap \operatorname{aff} C = H'$. (If already aff $C = \mathbb{R}^d$, then $H = H'$.) Then we also have $x \in H$ and $H \cap \operatorname{ri} C = \emptyset$. Theorem 4.1, (a) $\Rightarrow$ (b) finally shows that H is in fact a proper supporting hyperplane. □

The following theorem is also fundamental:

Theorem 4.5. *Let C be a non-empty closed convex set in $\mathbb{R}^d$. Then C is the intersection of its supporting halfspaces.*

PROOF. When $\dim C = 0$, the theorem is clearly true. When $C = \mathbb{R}^d$, there are no supporting halfspaces; hence, the theorem is also true in this case. So, let $\dim C \geq 1$, and let x be a point of $\mathbb{R}^d$ outside C; we shall prove that there is a supporting halfspace K of C such that $x \notin K$. If $x \notin \operatorname{aff} C$, there is a hyperplane H in $\mathbb{R}^d$ with $\operatorname{aff} C \subset H$ and $x \notin H$. The closed halfspace bounded by H which does not contain x then has the desired property. If $x \in \operatorname{aff} C$, let z be a relative interior point of C, cf. Theorem 3.1. Then $[z, x] \cap C$ is a closed segment $[z, u]$, where $u \in \operatorname{rb} C$ and $[z, u[$ is in $\operatorname{ri} C$, cf. Theorem 3.3. Now, by Theorem 4.3 there is a proper supporting hyperplane H of C such that $u \in H$. The supporting halfspace K bounded by H then has the desired property. In fact, suppose that $x \in K$. As we have $z \in \operatorname{ri} C$, it follows from Corollary 4.2 that $z \notin H$, whence $z \in \operatorname{int} K$. But then Theorem 3.3 shows that $]z, x[$ is in $\operatorname{int} K$, which is contradicted by the fact that the point u belonging to $]z, x[$ is in $H = \operatorname{bd} K$. □

One may say that Theorem 4.5 describes an "external representation" of a closed convex set. In the next section we shall meet an "internal representation" of a compact convex set.

EXERCISES

4.1. Let C_1 and C_2 be convex sets in $\mathbb{R}^d$. A hyperplane H in $\mathbb{R}^d$ is said to *separate* C_1 and C_2 if C_1 is contained in one of the two closed halfspaces bounded by H and C_2 is contained in the other closed halfspace bounded by H. Note that we allow $C_1 \subset H$ and $C_2 \subset H$. If at least one of the two sets C_1 and C_2 is not contained in H, then H is said to separate *properly*. Show that there exists a hyperplane H separating C_1 and C_2 properly if and only if $\operatorname{ri} C_1 \cap \operatorname{ri} C_2 = \emptyset$. (Hint: Consider the convex set $C := C_1 - C_2$. Use Exercise 3.2.)

4.2. Let C_1 and C_2 be convex sets in $\mathbb{R}^d$. A hyperplane $H(y, \alpha)$ is said to separate C_1 and C_2 *strongly* if for some $\varepsilon > 0$ both $H(y, \alpha - \varepsilon)$ and $H(y, \alpha + \varepsilon)$ separate C_1 and C_2, cf. Exercise 4.1. Show that there exists a hyperplane H separating C_1 and C_2 strongly if and only if $o \notin \operatorname{cl}(C_1 - C_2)$. Deduce, in particular, that if C_1 and C_2 are disjoint closed convex sets one of which is compact, then there is a strongly separating hyperplane.

§5. The Facial Structure of a Closed Convex Set

In this section we shall study certain "extreme" convex subsets of a closed convex set C, called the faces of C. We shall prove, among other things, that when the set C is compact, then it is the convex hull of its 0-dimensional faces. This is the "internal representation" mentioned in Section 4.

In the following, let C be a closed convex set in $\mathbb{R}^d$. A convex subset F of C is called a *face* of C if for any two distinct points $y, z \in C$ such that $]y, z[\cap F$ is non-empty, we actually have $[y, z] \subset F$. Note that in order to have $[y, z] \subset F$ it suffices by the convexity of F to have $y, z \in F$.

The subsets $\emptyset$ and C of C are both faces of C, called the *improper faces*; all other faces are called *proper faces*.

A point $x \in C$ is called an *extreme point* of C if $\{x\}$ is a face. This means, by definition, that x is not a relative interior point of any segment $[y, z]$ in C, or, equivalently, that $C \setminus \{x\}$ is again convex. The set of extreme points of C is denoted by ext C.

A face F of C is called a *k-face* if $\dim F = k$. Thus, the 0-faces are the extreme points. (Strictly speaking, $\{x\}$ is a face if and only if x is an extreme point.) A *facet* of C is a face F with $0 \leq \dim F = \dim C - 1$.

It is clear that the intersection of any set $\mathscr{A}$ of faces of C is again a face of C. Hence, there is a largest face of C contained in all the members of $\mathscr{A}$, namely, the intersection of all the members of $\mathscr{A}$. However, we can also conclude that there is a smallest face of C containing all the members of $\mathscr{A}$, namely, the intersection of all faces of C containing all the members of $\mathscr{A}$. (Note that C itself is such a face.) Denoting the set of all faces of C by $\mathscr{F}(C)$, we may express this by saying that the partially ordered set $(\mathscr{F}(C), \subset)$ is a complete lattice with the lattice operations

$$\inf \mathscr{A} := \bigcap \{F \in \mathscr{F}(C) \mid F \in \mathscr{A}\},$$

$$\sup \mathscr{A} := \bigcap \{G \in \mathscr{F}(C) \mid \forall F \in \mathscr{A} : F \subset G\}.$$

(For lattice-theoretic notions, see Appendix 1.) We shall call $(\mathscr{F}(C), \subset)$ the *face-lattice* of C. (The partially ordered set $(\mathscr{F}(C), \supset)$ is, of course, also a complete lattice. However, when speaking of the face-lattice of C we always mean $\mathscr{F}(C)$ equipped with $\subset$.)

When C is a closed convex set with $\dim C \geq 1$, then certain faces of C have a particular form: If H is a proper supporting hyperplane of C, cf. Corollary 4.2, then the set $F := H \cap C$ is a proper face of C. In fact, F is a non-empty proper subset of C by definition, and being the intersection of two convex sets it is also convex. To see that it has the face property, let y and z be two points of C such that $]y, z[\cap F$ is non-empty. Then $(1 - \lambda)y + \lambda z$ is in H for some $\lambda \in]0, 1[$. Now, there are u and α such that $H = H(u, \alpha)$ and $C \subset K(u, \alpha)$. We then have $\langle y, u \rangle \leq \alpha$, $\langle z, u \rangle \leq \alpha$ and $\langle (1 - \lambda)y + \lambda z, u \rangle = \alpha$, whence $\langle y, u \rangle = \langle z, u \rangle = \alpha$, i.e. y and z are in H, and therefore in F, as desired. A face F of C of the form $F = H \cap C$, where H is a proper supporting hyperplane of C, is called a (*proper*) *exposed face*. For any closed convex set C (including sets C with $\dim C = -1, 0$) it is convenient also to consider $\emptyset$ and C as exposed faces of C; we shall call them *improper exposed faces*.

(There is a formal problem in connection with the definition of a proper exposed face of C, namely, that it depends on the choice of the particular affine space containing C. If C is "initially" lying in $\mathbb{R}^d$, we would like to be free to consider it as a subset of any affine subspace A of $\mathbb{R}^d$ containing aff C. We can, however, easily get away with this difficulty, since the hyperplanes in A are just the non-empty intersections $H \cap A$, where H is a hyperplane in $\mathbb{R}^d$ not containing A.)

A point $x \in C$ is called an *exposed point* of C if $\{x\}$ is an exposed face. The set of exposed points of C is denoted by exp C. Thus, exp C is a subset of ext C.

The set of exposed faces of C is denoted by $\mathscr{E}(C)$. The order-theoretic structure of $(\mathscr{E}(C), \subset)$ will be discussed later in this section.

In order to illustrate the notions introduced above, consider the following example. Let C be the convex hull of two disjoint closed discs in $\mathbb{R}^2$ having the same radius. Then the boundary of C consists of two closed segments $[x_1, x_2]$ and $[x_3, x_4]$, and two open half-circles. The 1-faces of C are the two segments $[x_1, x_2]$, $[x_3, x_4]$; these faces are in fact exposed. The extreme points (i.e. the 0-faces) are the points x_1, x_2, x_3, x_4 and the points belonging to one of the open half-circles. Clearly, each point belonging to one of the open half-circles is even exposed. The extreme points x_1, x_2, x_3, x_4 are not exposed however; in fact, a supporting hyperplane of C containing one of the points x_1, x_2, x_3, x_4 must also contain one of the two segments. In particular, this shows that in general there are non-exposed faces.

Any proper exposed face is the intersection of two closed sets, and therefore it is closed itself. We actually have:

Theorem 5.1. *Every face F of a closed convex set C in $\mathbb{R}^d$ is closed.*

PROOF. For dim $F = -1, 0$ there is nothing to prove. Assume that dim $F \geq 1$, and let x be any point in cl F. Let x_0 be a point in ri F, cf. Theorem 3.1. If $x = x_0$, we have $x \in F$ as desired. If $x \neq x_0$, then $[x_0, x[$ is a subset of ri F by Theorem 3.3. In particular, $]x_0, x[\cap F \neq \emptyset$, whence x is in F by the definition of a face. □

Theorem 5.1 shows among other things that it makes sense to talk about "a face of a face" (of a closed convex set):

Theorem 5.2. *Let F be a face of a closed convex set C in $\mathbb{R}^d$, and let G be a subset of F. Then G is a face of C if (and only if) G is a face of F.*

PROOF. It follows immediately from the definition that if the set G is a face of C, then it is also a face of F. Conversely, suppose that G is a face of F, and let y and z be points of C such that $]y, z[$ intersects G. Since $G \subset F$, the segment $]y, z[$ also intersects F. This implies $y, z \in F$ since F is a face of C. But then we also have $y, z \in G$, as desired, since G is a face of F. □

One should note that the "if" part of Theorem 5.2 is not valid in general with "face" replaced everywhere by "exposed face." In fact, in the example above x_1 is an exposed point of $[x_1, x_2]$, and $[x_1, x_2]$ is an exposed face of C, but x_1 is not an exposed point of C.

Theorem 5.3. *Let F be a face of a closed convex set C in $\mathbb{R}^d$ such that $F \neq C$. Then $F \subset$ rb C.*

PROOF. For dim $C = -1, 0$ there is nothing to prove. So, assume that we have dim $C \geq 1$. Let F be a face of C such that some point x from F is in ri C. We shall complete the proof by showing that $F = C$. Let y be an arbitrary point in C. If $y = x$, then y is in F, as desired. If $y \neq x$, then there is a point z in C such that $x \in]y, z[$, cf. Theorem 3.5, (a) $\Rightarrow$ (c). Since x is in F, and F is a face, it follows that y is in F. □

Corollary 5.4. *Let F and G be faces of a closed convex set C in $\mathbb{R}^d$ such that $G \subsetneq F$. Then $G \subset$ rb F.*

PROOF. First note that G is a face of F, cf. Theorem 5.2. The statement then follows immediately from Theorem 5.3. □

Corollary 5.5. *Let F and G be faces of a closed convex set C in $\mathbb{R}^d$ such that $G \subsetneq F$. Then* dim $G <$ dim F.

PROOF. First note that we have aff $G \subset$ aff F since $G \subset F$. Suppose that aff $G =$ aff F. Then ri $G \subset$ ri F since $G \subset F$. Combining with Corollary 5.4 we obtain ri $G = \varnothing$. By Theorem 3.1 this implies $G = \varnothing$, whence also $F = \varnothing$ since aff $F =$ aff G, contradicting that $G \neq F$ by assumption. In conclusion, we must have aff $G \subsetneq$ aff F, whence dim $G <$ dim F. □

For any subset M of a closed convex set C in $\mathbb{R}^d$ there is a smallest face of C containing M, namely, the intersection of all faces containing M. Theorem 5.3 shows that when M contains a point from ri C, then the smallest face containing M is C itself.

Theorem 5.6. *Let C be a closed convex set in $\mathbb{R}^d$, let x be a point in C, and let F be a face of C containing x. Then F is the smallest face of C containing x if and only if $x \in$ ri F.*

PROOF. If $x \in$ ri F, then F is the smallest face containing x by Corollary 5.4. If $x \in$ rb F, then by Theorem 4.3 there is a face G (in fact, exposed) of F such that $x \in G \subsetneq F$. By Theorem 5.2, G is also a face of C, and therefore F is not the smallest face containing x. □

Corollary 5.7. *Let C be a closed convex set in $\mathbb{R}^d$. Then the sets* ri F, *where $F \in \mathscr{F}(C) \setminus \{\varnothing\}$, form a partition of C.*

PROOF. The statement amounts to saying that for each $x \in C$ there is a unique face F of C such that $x \in$ ri F. However, Theorem 5.6 gives such a unique face, namely, the smallest face of C containing x. □

Next, we shall study the exposed faces. We first prove:

Theorem 5.8. *Let F be a facet of a closed convex set C in $\mathbb{R}^d$. Then F is an exposed face.*

PROOF. By the definition of a facet we necessarily have $\dim F \geq 0$, whence by Theorem 3.1 there is a point $x \in \operatorname{ri} F$. Then, by Theorem 5.6, F is the smallest face of C containing x. On the other hand, Theorem 4.3 shows that there is an exposed face G of C such that $x \in G$. It then follows that $F \subset G \subsetneqq C$. Using Corollary 5.5 we obtain

$$\dim C - 1 = \dim F \leq \dim G < \dim C,$$

whence $\dim G = \dim F$. Corollary 5.5 then shows that $F = G$, and therefore F is exposed. □

At the beginning of this section we noted that the intersection of any set of faces of a closed convex set C is again a face of C. The following theorem shows that a similar result holds for exposed faces:

Theorem 5.9. *Let $\{F_i \mid i \in I\}$ be a set of exposed faces of a closed convex set C in $\mathbb{R}^d$, and let*

$$F := \bigcap_{i \in I} F_i.$$

Then F is also an exposed face of C.

PROOF. When F is $\emptyset$ or C, there is nothing to prove. So, in the following we may assume that F is a non-empty intersection of proper exposed faces F_i, $i \in I$.

We shall first consider the case where I is a finite set, say $I = \{1, \ldots, n\}$. Now, for each $i \in I$ there is a hyperplane $H(y_i, \alpha_i)$ such that

$$F_i = H(y_i, \alpha_i) \cap C \tag{1}$$

and

$$C \subset K(y_i, \alpha_i). \tag{2}$$

We may assume without loss of generality that $o \in \operatorname{int} C$. Then o is interior for all the $K(y_i, \alpha_i)$'s, and therefore each α_i is >0. Letting

$$z_i := \alpha_i^{-1} y_i$$

for $i = 1, \ldots, n$, (1) and (2) become

$$F_i = H(z_i, 1) \cap C,$$

$$C \subset K(z_i, 1).$$

Let

$$z_0 := \sum_{i=1}^{n} n^{-1} z_i.$$

Then for any $x \in C$ we have

$$\begin{aligned}\langle x, z_0 \rangle &= \left\langle x, \sum_{i=1}^{n} n^{-1} z_i \right\rangle \\ &= \sum_{i=1}^{n} n^{-1} \langle x, z_i \rangle \\ &\leq \sum_{i=1}^{n} n^{-1} \cdot 1 = 1, \end{aligned} \tag{3}$$

implying that

$$C \subset K(z_0, 1).$$

Furthermore, we have equality in (3) if and only if $x \in H(z_i, 1)$ for $i = 1, \ldots, n$. This shows that

$$H(z_0, 1) \cap C = F.$$

Hence, F is an exposed face.

When I is infinite, it suffices by the preceding to prove that there exist $i_1, \ldots, i_n \in I$ such that

$$\bigcap_{\nu=1}^{n} F_{i_\nu} = F.$$

Let i_1 be any of the i's in I. If $F = F_{i_1}$, we have the desired conclusion. If $F \subsetneqq F_{i_1}$, then there is $i_2 \in I$ such that

$$F \subset F_{i_1} \cap F_{i_2} \subsetneqq F_{i_1}.$$

From Corollary 5.5 it follows that

$$\dim(F_{i_1} \cap F_{i_2}) < \dim F_{i_1}.$$

If $F = F_{i_1} \cap F_{i_2}$, we have the desired conclusion. If $F \subsetneqq F_{i_1} \cap F_{i_2}$, then there is $i_3 \in I$ such that

$$F \subset F_{i_1} \cap F_{i_2} \cap F_{i_3} \subsetneqq F_{i_1} \cap F_{i_2}.$$

Again from Corollary 5.5 it follows that

$$\dim(F_{i_1} \cap F_{i_2} \cap F_{i_3}) < \dim(F_{i_1} \cap F_{i_2}).$$

If $F = F_{i_1} \cap F_{i_2} \cap F_{i_3}$, we have the desired conclusion. If $F \subsetneqq F_{i_1} \cap F_{i_2} \cap F_{i_3}$, there is $i_4 \in I$, etc. Since the dimension in each step is lowered by at least 1, we must end up with $i_1, \ldots, i_n \in I$ such that

$$F = \bigcap_{\nu=1}^{n} F_{i_\nu},$$

as desired. □

It follows from Theorem 5.9 that the partially ordered set $(\mathscr{E}(C), \subset)$ of exposed faces of a closed convex set C in $\mathbb{R}^d$ is a complete lattice with

$$\inf \mathscr{A} := \bigcap \{F \in \mathscr{E}(C) \mid F \in \mathscr{A}\}$$

$$\sup \mathscr{A} := \bigcap \{G \in \mathscr{E}(C) \mid \forall F \in \mathscr{A} : F \subset G\}$$

for $\mathscr{A} \subset \mathscr{E}(C)$. It is interesting to note, however, that in general $(\mathscr{E}(C), \subset)$ is not a sublattice (cf. Appendix 1) of $(\mathscr{F}(C), \subset)$. In fact, when $\mathscr{A}$ is a subset of $\mathscr{E}(C)$, then $\sup \mathscr{A}$ computed in $(\mathscr{E}(C), \subset)$ may be different from $\sup \mathscr{A}$ computed in $(\mathscr{F}(C), \subset)$. (The inf-operation, however, is the same in $(\mathscr{E}(C), \subset)$ as in $(\mathscr{F}(C), \subset)$.) For example, it is not difficult to construct in $\mathbb{R}^3$ a closed convex set C with the following properties. Among the extreme points of C there are three, say x_1, x_2, x_3, such that $\operatorname{conv}\{x_1, x_2, x_3\}$ is an exposed face, x_1 and x_2 are exposed points, but the face $[x_1, x_2]$ is not exposed. (See Figure 2.) Then if we consider the subset $\mathscr{A}$ of $\mathscr{E}(C)$ consisting of the two exposed faces $\{x_1\}$ and $\{x_2\}$, we see that $\sup \mathscr{A}$ in $(\mathscr{F}(C), \subset)$ is $[x_1, x_2]$, whereas $\sup \mathscr{A}$ in $(\mathscr{E}(C), \subset)$ is $\operatorname{conv}\{x_1, x_2, x_3\}$.

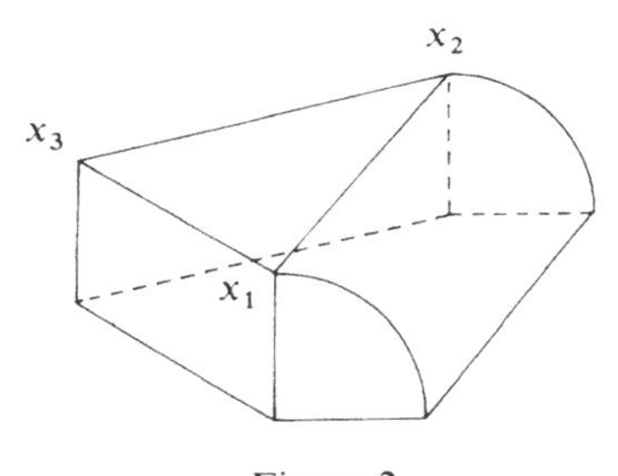

Figure 2

The final theorem of this section deals with extreme points. Closed half-spaces and affine subspaces are closed convex sets without extreme points. We shall prove that compact convex sets are "spanned" by their extreme points. This result is known as *Minkowski's Theorem*:

Theorem 5.10. *Let C be a compact convex set in $\mathbb{R}^d$, and let M be a subset of C. Then the following two conditions are equivalent:*

(a) $C = \operatorname{conv} M$.
(b) $\operatorname{ext} C \subset M$.

In particular,

(c) $C = \operatorname{conv}(\operatorname{ext} C)$.

Proof. Suppose that there is an extreme point x of C which is not in M. Then M is a subset of $C\setminus\{x\}$, and since $C\setminus\{x\}$ is convex by the definition of an extreme point, it follows that $\operatorname{conv} M$ is also a subset of $C\setminus\{x\}$. This proves (a) $\Rightarrow$ (b).

To prove (b) ⇒ (a) it suffices to show that

$$C \subset \operatorname{conv}(\operatorname{ext} C). \tag{4}$$

(In fact, suppose that (4) holds. Since the opposite inclusion of (4) is obvious, it then follows that $C = \operatorname{conv}(\operatorname{ext} C)$. But then we also have $C = \operatorname{conv} M$ for any subset M of C containing ext C.) We shall prove (4) by induction on the dimension of C. For dim $C = -1, 0$ there is nothing to prove. For dim $C = 1$ the statement is clearly valid. Suppose that the statement is valid for all compact convex sets of dimension $<e$, where $e \geq 2$, and let C be a compact convex set of dimension e. Let x be any point in C; we shall prove that x is a convex combination of extreme points of C, cf. Theorem 2.2. If x itself is an extreme point, there is nothing to prove. If x is not an extreme point, then there is a segment in C having x in its relative interior. Extending the segment, if necessary, we see that there are in fact points $y_0, y_1 \in \operatorname{rb} C$ such that $x \in]y_0, y_1[$. Let F_0 and F_1 be the smallest faces of C containing y_0 and y_1, respectively. Then F_0 and F_1 are proper faces of C, cf. Corollary 5.7. They are, in particular, compact convex sets, cf. Theorem 5.1, and they both have dimension $<e$, cf. Corollary 5.5. Then, by the induction hypothesis, there are points $x_{01}, \ldots, x_{0p} \in \operatorname{ext} F_0$ and $x_{11}, \ldots, x_{1q} \in \operatorname{ext} F_1$ such that y_0 is a convex combination of the x_{0i}'s and y_1 is a convex combination of the x_{1j}'s. Since x is a convex combination of y_0 and y_1, it follows that x is a convex combination of the x_{0i}'s and x_{1j}'s. To complete the proof, we note that the x_{0i}'s and x_{1j}'s are in fact extreme points of C; this follows from Theorem 5.2. □

Corollary 5.11. *Let C be a compact convex set in $\mathbb{R}^d$ with* dim $C = n$. *Then each point of C is a convex combination of at most $n + 1$ extreme points of C.*

Proof. Combine Theorem 5.10(c) and Corollary 2.4. □

Exercises

5.1. Show that ext C is closed when C is a 2-dimensional compact convex set.

5.2. Let C be the convex hull of the set of points $(\alpha_1, \alpha_2, \alpha_3) \in \mathbb{R}^3$ such that

$$\alpha_1 = \alpha_2 = 0, \qquad \alpha_3 \in [-1, 1],$$

or

$$\alpha_3 = 0, \qquad (\alpha_1 - 1)^2 + \alpha_2^2 = 1.$$

Show that ext C is non-closed.

5.3. Let C be a closed convex set in $\mathbb{R}^d$. Show that if a convex subset F of C is a face of C, then $C \setminus F$ is convex. Show that the converse does not hold in general.

5.4. Let C be a non-empty closed convex set in $\mathbb{R}^d$. An affine subspace A of $\mathbb{R}^d$ is said to *support* C if $A \cap C \neq \emptyset$ and $C \setminus A$ is convex. Show that the supporting hyperplanes of C in the sense of Section 4 are the hyperplanes that support C in the sense just defined. Show that for a non-empty convex subset F of C, the following three conditions are equivalent:

(a) F is a face of C.
(b) There is a supporting affine subspace A of C such that $A \cap C = F$.
(c) aff F is a supporting affine subspace of C with $(\text{aff } F) \cap C = F$.

(The equivalence of (a) and (b) throws some light upon the difference between faces and exposed faces.)

5.5. Let C be a compact convex set in $\mathbb{R}^d$, and let M be a subset of ext C. Show that conv M is a face of C if and only if

$$(\text{aff } M) \cap \text{conv}((\text{ext } C)\backslash M) = \emptyset.$$

5.6. Show that there are compact convex sets C such that

$$C \neq \text{conv}(\exp C).$$

Prove *Straszewicz's Theorem*: For any compact convex set C one has

$$C = \text{clconv}(\exp C).$$

(Warning: This is not easy.)

§6. Polarity

Duality plays an important role in convexity theory in general, and in polytope theory in particular. Actually, we shall be working with two duality concepts: a narrow one called polarity and a broader one which we shall simply refer to as duality. The notion of polarity applies to convex sets in general, whereas duality in the broader sense will only be applied to polytopes.

This section deals with polarity. With each subset M of $\mathbb{R}^d$, we shall associate a certain closed convex subset M° of $\mathbb{R}^d$, called the polar of M. When C is a compact convex set having o in its interior, then the polar set C° has the same properties, and C is the polar of C°. For such a pair of mutually polar compact convex sets having o as interior point, the polar operation induces a one-to-one inclusion reversing correspondence between $\mathscr{F}(C)$ and $\mathscr{F}(C^\circ)$.

One should note that the notion of polarity is a linear concept, while in the preceding Sections 2–5 we worked within the framework of affine spaces. In particular, the polar operation is not translation invariant.

For any subset M of $\mathbb{R}^d$, the *polar set* is the subset M° of $\mathbb{R}^d$ defined by

$$\begin{aligned} M^\circ &:= \{y \in \mathbb{R}^d \mid \forall x \in M : \langle x, y \rangle \leq 1\} \\ &= \{y \in \mathbb{R}^d \mid \sup_{x \in M} \langle x, y \rangle \leq 1\}. \end{aligned}$$

Equivalently,

$$M^\circ = \bigcap_{x \in M} K(x, 1). \tag{1}$$

Since y is in $K(x, 1)$ if and only if x is in $K(y, 1)$, it follows from (1) that we have

$$y \in M^\circ \Leftrightarrow M \subset K(y, 1). \tag{2}$$

It also follows from (1) that M° is a closed convex set containing o, since each $K(x, 1)$ is such a set. Furthermore, it is clear that

$$M_1 \subset M_2 \Rightarrow M_1^\circ \supset M_2^\circ. \tag{3}$$

We shall prove the following:

Theorem 6.1. *For any subset M of $\mathbb{R}^d$ one has:*

(a) *If M is bounded, then o is an interior point of M°.*
(b) *If o is an interior point of M, then M° is bounded.*

Proof. For $z \in \mathbb{R}^d$ and $r > 0$ we denote by $B(z, r)$ the closed ball centred at z with radius r, i.e.

$$B(z, r) := \{x \in \mathbb{R}^d \mid \|x - z\| \leq r\}.$$

Here $\|\cdot\|$ denotes the Euclidean norm, i.e.

$$\|u\| = \sqrt{\langle u, u \rangle}.$$

Now, it is an elementary standard fact that

$$\sup_{x \in B(o, r)} \langle x, y \rangle = r\|y\|$$

for all $y \in \mathbb{R}^d$ and $r > 0$. This shows that

$$B(o, r)^\circ = B(o, r^{-1}). \tag{4}$$

Therefore, if M is bounded, i.e. $M \subset B(o, r)$ for some $r > 0$, then using (3) and (4) we see that $B(o, r^{-1}) \subset M^\circ$, showing that o is an interior point of M°. This proves (a). Next, if o is an interior point of M, i.e. $B(o, r) \subset M$ for some $r > 0$, then again using (3) and (4) we obtain $M^\circ \subset B(o, r^{-1})$, showing that M° is bounded. This proves (b). □

The polar operation can, of course, be iterated. We write $M^{\circ\circ}$ instead of $(M^\circ)^\circ$. The set $M^{\circ\circ}$ is called the *bipolar* of M. It can be described as follows:

Theorem 6.2. *For any subset M of $\mathbb{R}^d$ we have*

$$M^{\circ\circ} = \operatorname{clconv}(\{o\} \cup M),$$

i.e. $M^{\circ\circ}$ is the smallest closed convex set containing o and M.

Proof. We have

$$M^{\circ\circ} = \bigcap_{y \in M^\circ} K(y, 1) = \bigcap_{M \subset K(y, 1)} K(y, 1), \tag{5}$$

cf. (1) and (2). This formula immediately implies that $M^{\circ\circ}$ is a closed convex set containing o and M, whence $M^{\circ\circ}$ contains $\operatorname{clconv}(\{o\} \cup M)$. To prove the opposite inclusion, let z be a point not in $\operatorname{clconv}(\{o\} \cup M)$; we shall prove that there is a closed halfspace $K(u, 1)$ containing M such that $z \notin K(u, 1)$, cf. (5). By Theorem 4.5 there is a supporting halfspace $K(y, \alpha)$ of $\operatorname{clconv}(\{o\} \cup M)$ such that $z \notin K(y, \alpha)$. We then have

$$\max\{\langle x, y\rangle \mid x \in \operatorname{clconv}(\{o\} \cup M)\} = \alpha < \langle z, y\rangle.$$

Since o is in $\operatorname{clconv}(\{o\} \cup M)$, we have $\alpha \geq 0$. Therefore, there exists $\beta > 0$ such that

$$\max\{\langle x, y\rangle \mid x \in \operatorname{clconv}(\{o\} \cup M)\} \leq \beta < \langle z, y\rangle. \tag{6}$$

Taking $u := \beta^{-1}y$, we obtain from (6)

$$\max\{\langle x, u\rangle \mid x \in \operatorname{clconv}(\{o\} \cup M)\} \leq 1 < \langle z, u\rangle,$$

implying $M \subset K(u, 1)$ and $z \notin K(u, 1)$, as desired. □

From Theorems 6.1 and 6.2 we immediately get:

Corollary 6.3. *Let C be a compact convex set in $\mathbb{R}^d$ having o as an interior point. Then C° is also a compact convex set having o as an interior point. Furthermore, $C^{\circ\circ} = C$.*

In the following, C is assumed to be a compact convex set in $\mathbb{R}^d$ with $o \in \operatorname{int} C$. To emphasize the completely symmetric roles played by C and C°, as explained by Corollary 6.3, we denote C° by D.

The assumption $o \in \operatorname{int} C$ implies that every supporting hyperplane of C is a proper supporting hyperplane, and has the form $H(y, 1)$ for a unique $y \in \mathbb{R}^d \setminus \{o\}$. We then have $C \subset K(y, 1)$, and hence $y \in D$. The following theorem gives more information about this situation:

Theorem 6.4. *For any $y \in \mathbb{R}^d$, the following two conditions are equivalent:*

(a) *$H(y, 1)$ is a supporting hyperplane of C.*
(b) *$y \in \operatorname{bd} D$.*

Similarly, for any $x \in \mathbb{R}^d$, the following two conditions are equivalent:

(c) *$H(x, 1)$ is a supporting hyperplane of D.*
(d) *$x \in \operatorname{bd} C$.*

PROOF. If (a) holds, then $y \in D$ and

$$\sup_{x \in C} \langle x, y\rangle = 1. \tag{7}$$

(Actually, the supremum is a maximum.) If we had $y \in \operatorname{int} D$, then we would also have $\lambda y \in D$ for a certain $\lambda > 1$. Since D is the polar of C, we would then have

$$\sup_{x \in C} \langle x, \lambda y\rangle \leq 1,$$

contradicting (7). Hence $y \in D \setminus \operatorname{int} D = \operatorname{bd} D$, as desired.

Conversely, if (b) holds, then, in particular, y is in $D \setminus \{o\}$. Since D is the polar of C, we then have

$$0 < \sup_{x \in C} \langle x, y \rangle \leq 1. \tag{8}$$

Now, if in (8) we had <1, then we would have

$$\sup_{x \in C} \langle x, \lambda y \rangle = 1$$

for a suitable $\lambda > 1$, whence λy would be in C° $(=D)$. Since $o \in \operatorname{int} D$ and $y \in]o, \lambda y[$, this would imply $y \in \operatorname{int} D$ by Theorem 3.3, a contradiction. Therefore,

$$\sup_{x \in C} \langle x, y \rangle = 1.$$

Finally, this supremum is actually a maximum by the compactness of C and the continuity of $\langle \cdot, y \rangle$. Hence, $H(y, 1)$ is a supporting hyperplane of C, as desired.

As explained earlier, C and D play completely symmetric roles. Therefore, the equivalence of (c) and (d) is a consequence of the equivalence of (a) and (b). □

Corollary 6.5. *For any $x, y \in \mathbb{R}^d$, the following four conditions are equivalent:*

(a) *$H(y, 1)$ is a supporting hyperplane of C at x.*
(b) *$H(x, 1)$ is a supporting hyperplane of D at y.*
(c) *$\langle x, y \rangle = 1$, $x \in \operatorname{bd} C$, $y \in \operatorname{bd} D$.*
(d) *$\langle x, y \rangle = 1$, $x \in C$, $y \in D$.*

PROOF. The equivalence (a) $\Leftrightarrow$ (c) follows immediately from Theorem 6.4, (a) $\Leftrightarrow$ (b). The equivalence (b) $\Leftrightarrow$ (c) then follows by symmetry, or from Theorem 6.4, (c) $\Leftrightarrow$ (d). It is trivial that (c) $\Rightarrow$ (d). We shall complete the proof by showing that (d) $\Rightarrow$ (a). From $y \in D$ $(= C^\circ)$ it follows that $C \subset K(y, 1)$, and from $\langle x, y \rangle = 1$ it follows that $x \in H(y, 1)$. Since $x \in C$, it then follows that $H(y, 1)$ is a supporting hyperplane of C at x. □

Now, for an exposed face F of C, proper or improper, we define

$$F^\triangle := \{y \in D \mid \forall x \in F : \langle x, y \rangle = 1\}.$$

Similarly, for an exposed face G of D we define

$$G^\triangle := \{x \in C \mid \forall y \in G : \langle x, y \rangle = 1\}.$$

The motivation for this concept is the fact that when F is a proper exposed face of C, then a point $y \in \mathbb{R}^d$ is in $F^\triangle$ if and only if $H(y, 1)$ is a supporting hyperplane of C with $F \subset H(y, 1)$; this follows immediately from Corollary 6.5, (a) $\Leftrightarrow$ (d). The same holds for a proper exposed face G of D. For the improper exposed faces C and $\emptyset$ of C, we have $C^\triangle = \emptyset$ and $\emptyset^\triangle = D$. And for the improper exposed faces D and $\varnothing$ of D, we have $D^{\triangle} = \varnothing$ and $\varnothing^{\triangle} = C$. (The unpleasant feature that $\varnothing^{\triangle}$ may have different "values" is, of course, due to the fact that we use the same notation for different mappings.)

Theorem 6.6. *Let F be a proper exposed face of C. Then $F^{\triangle}$ is a proper exposed face of D. Similarly for a proper exposed face G of D.*

PROOF. By definition,

$$F^{\triangle} = \bigcap_{x \in F} D \cap H(x, 1).$$

When F is proper, then each $x \in F$ is in bd C, whence $H(x, 1)$ is a supporting hyperplane of D, cf. Theorem 6.4, (d) $\Rightarrow$ (c). Therefore, each set $D \cap H(x, 1)$ is a proper exposed face of D, implying that $F^{\triangle}$ is an exposed face of D, cf. Theorem 5.9. Furthermore, $F^{\triangle}$ is proper or empty. But since F is a proper exposed face, there is a supporting hyperplane $H(y, 1)$ of C such that $F = C \cap H(y, 1)$. From the remark above following the definition of $F^{\triangle}$, we then see that $y \in F^{\triangle}$, whence $F^{\triangle} \neq \varnothing$. □

By Theorem 6.6, it makes sense to iterate the $\triangle$-operation. Writing $F^{\triangle\triangle}$ instead of $(F^{\triangle})^{\triangle}$, we see that for the improper exposed faces C and $\varnothing$ of C we have $C^{\triangle\triangle} = C$ and $\varnothing^{\triangle\triangle} = \varnothing$. Moreover, by Theorem 6.6, $F^{\triangle\triangle}$ is a proper exposed face of C when F is a proper exposed face of C. We actually have:

Theorem 6.7. *Let F be a proper exposed face of C. Then $F^{\triangle\triangle} = F$. Similarly for a proper exposed face G of D.*

PROOF. By definition,

$$F^{\triangle\triangle} = \bigcap_{y \in F^{\triangle}} C \cap H(y, 1).$$

But since y is in $F^{\triangle}$ if and only if $H(y, 1)$ is a supporting hyperplane of C with $F \subset H(y, 1)$, we see that $F^{\triangle\triangle}$ is the intersection of all proper exposed faces of C containing F. This intersection, of course, is simply F itself. □

For an exposed face F of C, we call the exposed face $F^{\triangle}$ of D the *conjugate face* of F; the same applies to an exposed face G of D. Theorems 6.6 and 6.7 show that the exposed faces of C and D go together in pairs F, G of mutually conjugate faces, both proper or both improper.

It is clear that the $\triangle$-operation reverses inclusions. The following is, therefore, a consequence of Theorems 6.6 and 6.7:

Corollary 6.8. *The mapping $F \mapsto F^{\triangle}$, where $F \in \mathscr{E}(C)$, is an anti-isomorphism from $(\mathscr{E}(C), \subset)$ onto $(\mathscr{E}(D), \subset)$, and the mapping $G \mapsto G^{\triangle}$, where $G \in \mathscr{E}(D)$, is an anti-isomorphism from $(\mathscr{E}(D), \subset)$ onto $(\mathscr{E}(C), \subset)$. The two mappings are mutually inverse.*

Anti-isomorphisms reverse inf and sup. Therefore, Corollary 6.8 yields:

Corollary 6.9. *Let $\{F_i \mid i \in I\}$ be a set of exposed faces of C, let F_0 denote the largest exposed face of C contained in all the F_i's (i.e. F_0 is the intersection of the F_i's), and let F_1 denote the smallest exposed face of C containing all the F_i's. Then $F_0^{\triangle}$ is the smallest exposed face of D containing all the $F_i^{\triangle}$'s, and $F_1^{\triangle}$ is the largest exposed face of D contained in all the $F_i^{\triangle}$'s (i.e. $F_1^{\triangle}$ is the intersection of the $F_i^{\triangle}$'s). Similarly for a set of exposed faces of D.*

We remind the reader that for some time we have been working under the general assumption that C and D are mutually polar compact convex sets in $\mathbb{R}^d$ having o as an interior point. This assumption is maintained in the following theorem. (Among other things, this explains the meaning of d in the formula.)

Theorem 6.10. *Let F and G be a pair of mutually conjugate faces of C and D, respectively. Then*

$$\dim F + \dim G \leq d - 1.$$

PROOF. The conjugate face of the improper exposed face $\emptyset$ of C is the improper exposed face D of D. Similarly, the conjugate face of the improper exposed face C of C is the improper exposed face $\emptyset$ of D. Since $\dim \emptyset = -1$, $\dim C = d$ and $\dim D = d$, we see that the formula holds when F is improper, in fact with equality. Consequently, we need only consider the case where F is a proper exposed face of C; then the conjugate face G of D is also proper, cf. Theorem 6.6. Now, by the definition of the $\triangle$-operation,

$$G = D \cap \bigcap_{x \in F} H(x, 1).$$

Therefore, G is a subset of the affine subspace $\bigcap_{x \in F} H(x, 1)$, whence

$$\dim G \leq \dim \bigcap_{x \in F} H(x, 1). \tag{9}$$

By (9), the affine subspace $\bigcap_{x \in F} H(x, 1)$ is non-empty; therefore it is a translate of the linear subspace $\bigcap_{x \in F} H(x, 0)$, and so

$$\dim \bigcap_{x \in F} H(x, 1) = \dim \bigcap_{x \in F} H(x, 0). \tag{10}$$

But

$$\begin{aligned} \bigcap_{x \in F} H(x, 0) &= \{y \in \mathbb{R}^d \mid \forall x \in F \colon \langle x, y \rangle = 0\} \\ &= F^{\perp} = (\operatorname{span} F)^{\perp}. \end{aligned}$$

Therefore,

$$\begin{aligned}\dim \bigcap_{x \in F} H(x, 0) &= \dim((\operatorname{span} F)^{\perp}) \\ &= d - \dim(\operatorname{span} F) \\ &= d - (\dim(\operatorname{aff} F) + 1) \\ &= d - 1 - \dim(\operatorname{aff} F) \\ &= d - 1 - \dim F, \end{aligned} \tag{11}$$

where we have used the fact that $o \notin \operatorname{aff} F$ to obtain

$$\dim(\operatorname{span} F) = \dim(\operatorname{aff} F) + 1.$$

Combining now (9), (10) and (11), we obtain the desired formula. □

Exercises

6.1. Show that $(\lambda M)^\circ = \lambda^{-1} M^\circ$ when $\lambda \neq 0$.

6.2. Show that $(M^{\circ\circ})^\circ = M^\circ$.

6.3. Show that $(\bigcup_{i \in I} M_i)^\circ = \bigcap_{i \in I} M_i^\circ$.

6.4. Show that

$$\left(\bigcap_{i \in I} C_i\right)^\circ = \operatorname{clconv} \bigcup_{i \in I} C_i^\circ$$

when the sets C_i are closed convex sets containing o.

6.5. For $e < d$, identify $\mathbb{R}^e$ with the subspace of $\mathbb{R}^d$ consisting of all $(x_1, \ldots, x_d) \in \mathbb{R}^d$ such that $x_{e+1} = \cdots = x_d = 0$. Let Π denote the orthogonal projection of $\mathbb{R}^d$ onto $\mathbb{R}^e$. Show that for any subset M of $\mathbb{R}^d$ we have

$$\Pi(M)^\circ = M^\circ \cap \mathbb{R}^e,$$

where $\Pi(M)^\circ$ denotes the polar of $\Pi(M)$ in $\mathbb{R}^e$ and M° denotes the polar of M in $\mathbb{R}^d$.

6.6. Let C and D be mutually polar compact convex sets. Let F be a proper exposed face of C, and let $G := F^\triangle$. Show that

$$G = D \cap \bigcap_{x \in \operatorname{ext} F} H(x, 1),$$

and show that

$$G = D \cap H(x_0, 1)$$

for any relative interior point x_0 of F.

6.7. Let C and D be mutually polar compact convex sets. Extend the definition of the $\triangle$-operation by allowing it to operate on arbitrary subsets of C and D. Show that when M is a subset of C, then $M^{\triangle\triangle} := (M^\triangle)^\triangle$ is the smallest exposed face of C containing M.

CHAPTER 2

Convex Polytopes

§7. Polytopes

A (convex) polytope is a set which is the convex hull of a non-empty finite set, see Section 2. We already know that polytopes are compact. We may, therefore, apply Section 5 on the facial structure of closed convex sets to polytopes. As one might expect, the facial structure of polytopes is considerably simpler than that of convex sets in general.

A polytope $P = \operatorname{conv}\{x_1, \ldots, x_n\}$ is called a *k-polytope* if $\dim P = k$. This means that some $(k + 1)$-subfamily of $(x_1, \ldots, x_n)$ is affinely independent, but no such $(k + 2)$-subfamily is affinely independent. By a *k-simplex* we mean a k-polytope which is a simplex. A simplex is a k-simplex if and only if it has $k + 1$ vertices, cf. Section 2. A 1-simplex is a closed segment. A 2-simplex is called a triangle, a 3-simplex is called a tetrahedron.

We have the following description of polytopes in terms of extreme points:

Theorem 7.1. *Let P be a non-empty subset of $\mathbb{R}^d$. Then the following two conditions are equivalent*:

(a) *P is a polytope.*
(b) *P is a compact convex set with a finite number of extreme points.*

Proof. When P is a polytope, say $P = \operatorname{conv}\{x_1, \ldots, x_n\}$, then P is compact by Corollary 2.9. Next, Theorem 5.10, (a) $\Rightarrow$ (b) shows that $\operatorname{ext} P$ is a subset of $\{x_1, \ldots, x_n\}$, and hence is a finite set. The converse follows immediately from Theorem 5.10, (b) $\Rightarrow$ (a). □

Following common usage, we shall henceforth call the extreme points, i.e. the 0-faces, of a polytope P the *vertices* of P. We shall continue to denote the set of vertices of P by ext P. The 1-faces are called the *edges* of P.

The vertices of a simplex S in the sense used in Section 2 are, in fact, the extreme points (i.e. vertices) of S. This follows immediately from Theorem 5.10 or Theorem 7.2 below.

The set $\{x_1, \ldots, x_n\}$ spanning a polytope $P = \text{conv}\{x_1, \ldots, x_n\}$ is of course not unique (except when P is a 1-point set); in fact, one may always add new points $x_{n+1}, \ldots$ already in P. However, there is a unique *minimal* spanning set, namely, the set ext P of vertices of P:

Theorem 7.2. *Let P be a polytope in $\mathbb{R}^d$, and let $\{x_1, \ldots, x_n\}$ be a finite subset of P. Then the following two conditions are equivalent:*

(a) $P = \text{conv}\{x_1, \ldots, x_n\}$.
(b) $\text{ext}\, P \subset \{x_1, \ldots, x_n\}$.

In particular,

(c) $P = \text{conv}(\text{ext}\, P)$.

PROOF. Noting that polytopes are compact, the statement follows immediately from Theorem 5.10. □

We shall next study the facial structure of polytopes in general.

Theorem 7.3. *Let P be a polytope in $\mathbb{R}^d$, and let F be a proper face of P. Then F is also a polytope, and* $\text{ext}\, F = F \cap \text{ext}\, P$.

PROOF. We begin by noting that P and F are compact, cf. Theorem 7.1, (a)⇒(b) and Theorem 5.1. Now, Theorem 5.2 shows that the extreme points of F are just those extreme points (vertices) of P which are in F, i.e. $\text{ext}\, F = F \cap \text{ext}\, P$. Since ext P is a finite set by Theorem 7.1, (a) ⇒ (b), it follows that ext F is a finite set. Application of Theorem 7.1, (b) ⇒ (a) completes the proof. □

Corollary 7.4. *Let P be a polytope in $\mathbb{R}^d$. Then the number of faces of P is finite.*

PROOF. The number of extreme points of P is finite by Theorem 7.2, (a) ⇒ (b). Each face of P is the convex hull of extreme points of P by Theorem 7.3 and Theorem 7.2(c). Therefore, the number of faces is finite. □

The following is a main result:

Theorem 7.5. *Let P be a polytope in $\mathbb{R}^d$. Then every face of P is an exposed face.*

PROOF. It suffices to prove the statement for d-polytopes in $\mathbb{R}^d$. We shall use induction on d. For $d = 0$ there is nothing to prove, for $d = 1$ the statement is trivial, and for $d = 2$ it is obvious. Suppose that the statement is valid for all polytopes of dimension $<d$, where $d \geq 3$, and let P be a d-polytope in $\mathbb{R}^d$. For improper faces of P there is nothing to prove, so let F be a proper face of P. Let x be a relative interior point of F, cf. Theorem 3.1, and let H be a proper supporting hyperplane of P at x, cf. Theorem 4.3. Then $H \cap P$ is a proper exposed face of P containing x. Using Theorem 5.6, we see that $F \subset H \cap P$. If $F = H \cap P$, then F is exposed, as desired. If $F \subsetneqq H \cap P$, then F is a proper face of $H \cap P$, cf. Theorem 5.2. (See Figure 3.) Since

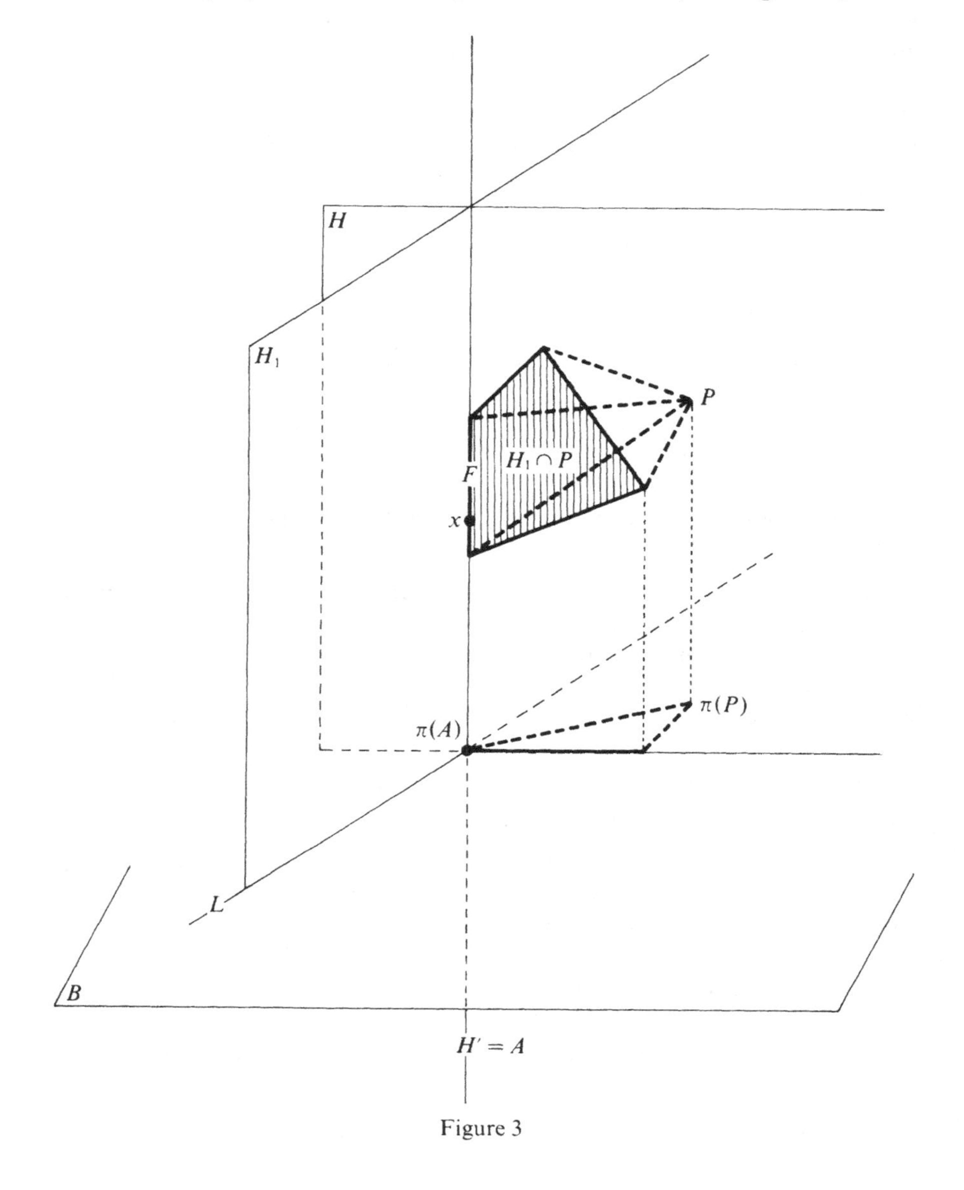

Figure 3

$\dim(H \cap P) < d$, and $H \cap P$ is a polytope, cf. Theorem 7.3, it follows from the induction hypothesis that there is a proper supporting hyperplane H' of $H \cap P$ in $\operatorname{aff}(H \cap P)$ such that $F = H' \cap (H \cap P)$. This hyperplane H' we may extend to a hyperplane A in H such that

$$F = A \cap P. \tag{1}$$

Note that $\dim A = d - 2 \geq 1$. Let B be a 2-dimensional affine subspace of $\mathbb{R}^d$ which is orthogonal to A, and let π denote the orthogonal projection of $\mathbb{R}^d$ onto B. Then $\pi(A)$ is a 1-point set. Furthermore, $\pi(P)$ is a 2-polytope in B. We claim that $\pi(A)$ is a vertex of $\pi(P)$. If not, then there are points y and z in P such that $\pi(y) \neq \pi(z)$ and

$$\pi(A) = (1 - \lambda)\pi(y) + \lambda\pi(z)$$

for some $\lambda \in]0, 1[$. Let

$$u := (1 - \lambda)y + \lambda z.$$

Then u is in P, and $\pi(u) = \pi(A)$, whence u is in A since $\pi^{-1}(\pi(A)) = A$. Therefore, u is in F, cf. (1). Since F is a face of P, it follows that y and z are in F. But F is a subset of A, whence $\pi(v) = \pi(A)$ for all $v \in F$. In particular, $\pi(y) = \pi(z)$, a contradiction which proves that $\pi(A)$ is a vertex of $\pi(P)$. By the 2-dimensional version of the theorem we then see that there is a line L in B such that

$$L \cap \pi(P) = \pi(A).$$

Then

$$H_1 := \operatorname{aff}(A \cup L) = \pi^{-1}(L)$$

is a supporting hyperplane of P in $\mathbb{R}^d$ with $H_1 \cap P = F$, as desired. □

Corollary 7.6. *Let P be a polytope in $\mathbb{R}^d$. Then the two lattices $(\mathscr{F}(P), \subset)$ and $(\mathscr{E}(P), \subset)$ are the same.*

We shall finally introduce two particular classes of polytopes, the pyramids and the bipyramids, and we shall describe their facial structure.

A *pyramid* in $\mathbb{R}^d$ is a polytope—cf. Theorem 7.7(a)—of the form

$$P = \operatorname{conv}(Q \cup \{x_0\}),$$

where Q is a polytope in $\mathbb{R}^d$, called the *basis* of P, and x_0 is a point of $\mathbb{R}^d \setminus \operatorname{aff} Q$, called the *apex* of P. (Note that basis and apex need not be unique: a simplex is a pyramid where any facet may be taken as the basis, or, equivalently, any vertex may be taken as the apex.) A pyramid P is called an *e-pyramid* if $\dim P = e$. Clearly, a pyramid P is an e-pyramid if and only if its basis Q is an $(e - 1)$-polytope.

The facial structure of a pyramid is determined by the facial structure of its basis as follows:

Theorem 7.7. *Let P be a pyramid in $\mathbb{R}^d$ with basis Q and apex x_0. Then the following holds:*

(a) *P is a polytope with* $\operatorname{ext} P = (\operatorname{ext} Q) \cup \{x_0\}$.
(b) *A subset F of P with $x_0 \notin F$ is a face of P if and only if it is a face of Q.*
(c) *A subset F of P with $x_0 \in F$ is a face of P if and only if there is a face G of Q such that* $F = \operatorname{conv}(G \cup \{x_0\})$, *i.e.* $F = \{x_0\}$ *or F is a pyramid with a face G of Q as the basis and x_0 as the apex. For each such face F of P, the face G is unique, and* $\dim G = \dim F - 1$.

PROOF. (a) The set

$$P_1 := \operatorname{conv}((\operatorname{ext} Q) \cup \{x_0\})$$

is a convex set containing Q and x_0, cf. Theorem 7.2(c). Therefore, it contains P. On the other hand, it is clear that $P_1 \subset P$, whence

$$P = \operatorname{conv}((\operatorname{ext} Q) \cup \{x_0\}).$$

This shows that P is a polytope and also implies that

$$\operatorname{ext} P \subset (\operatorname{ext} Q) \cup \{x_0\},$$

cf. Theorem 7.2, (a) $\Rightarrow$ (b). To prove the opposite inclusion, we first remark that P is the union of all segments $[y, x_0]$, where $y \in Q$. It is then clear that if H_0 is a hyperplane with $x_0 \in H_0$ and $H_0 \cap \operatorname{aff} Q = \emptyset$, then H_0 is a supporting hyperplane of P with $H_0 \cap P = \{x_0\}$, implying that $x_0 \in \operatorname{ext} P$. To prove also that every $x \in \operatorname{ext} Q$ is in $\operatorname{ext} P$, we prove more generally that every proper face of Q is a face of P. Let F be a proper face of Q. Then there is a supporting hyperplane H of Q in $\operatorname{aff} Q$ such that $H \cap Q = F$. Let H_1 be a hyperplane in $\mathbb{R}^d$ such that $H_1 \cap \operatorname{aff} Q = H$ and x_0 is on the same side of H_1 as $Q \setminus F$. Then, again using the remark above that each point of P belongs to some segment $[y, x_0]$ with $y \in Q$, we see that $H_1 \cap P = F$, whence F is a face of P. This completes the proof of (a). (A more direct way of showing that every (proper) face of Q is a face of P goes via the observation that Q is a facet of P. Our motivation for preferring the proof given above is the fact that after an obvious modification it also applies to the situation needed in the proof of Theorem 7.8 below.)

(b) During the proof of (a) it was proved that every proper face of Q is a face of P. Since Q itself is also a face (in fact, a facet) of P, it follows that every face of Q is a face of P.

Conversely, let F be a non-empty face of P not containing x_0. By Theorem 7.5 there is a supporting hyperplane H of P such that $H \cap P = F$. Using (a) and Theorem 7.3 we see that $\operatorname{ext} F \subset \operatorname{ext} Q$, whence $F \subset Q$. But then trivially F is a face of Q.

(c) We first prove that every set F of the form

$$F = \operatorname{conv}(G \cup \{x_0\}),$$

where G is a face of Q, is a face of P. We need only consider the case where G is a proper face of Q. For any such face G there is a supporting hyperplane H of Q in aff Q such that $H \cap Q = G$. Let H_1 be a hyperplane in $\mathbb{R}^d$ such that $H_1 \cap \operatorname{aff} Q = H$ and $x_0 \in H_1$. Then clearly H_1 is a proper supporting hyperplane of P, whence

$$F_1 := H_1 \cap P$$

is a proper (exposed) face of P. Moreover,

$$\begin{aligned} \operatorname{ext} F_1 &= H_1 \cap \operatorname{ext} P \\ &= (H \cap \operatorname{ext} Q) \cup \{x_0\} \\ &= (\operatorname{ext} G) \cup \{x_0\}, \end{aligned}$$

where we have used Theorem 7.3 and (a). Then using Theorem 7.2(c) we get

$$\begin{aligned} F_1 &= \operatorname{conv}((\operatorname{ext} G) \cup \{x_0\}) \\ &= \operatorname{conv}(G \cup \{x_0\}) \\ &= F, \end{aligned}$$

whence F is a face of P, as desired.

To prove the converse, we need only consider the case where $F \neq \{x_0\}$ and $F \neq P$. Let H be a supporting hyperplane of P such that $F = H \cap P$, cf. Theorem 7.5. Since P is the union of all segments $[y, x_0]$, where $y \in Q$, we see that F is the union of all segments $[y, x_0]$, where $y \in H \cap Q$. Letting $G := H \cap Q$, it follows that

$$F = \operatorname{conv}(G \cup \{x_0\}),$$

and it is clear that G is a face of Q.

Finally, the uniqueness of G and the dimension formula are obvious. □

A *bipyramid* in $\mathbb{R}^d$ is a polytope—cf. Theorem 7.8(a)—of the form

$$P = \operatorname{conv}(Q \cup \{x_0, x_1\}),$$

where Q is a polytope in $\mathbb{R}^d$ with $\dim Q \geq 1$, and x_0, x_1 are two points of $\mathbb{R}^d \setminus \operatorname{aff} Q$ such that

$$]x_0, x_1[\cap \operatorname{ri} Q \neq \varnothing.$$

(Then actually $]x_0, x_1[$ has precisely one point in common with ri Q.) The set Q is called the *basis* of P, and x_0, x_1 are called the *apices* of P. (As in the case of pyramids, basis and apices are, in general, not unique.) A bipyramid P is called an *e-bipyramid* if $\dim P = e$. Clearly, a bipyramid P is an e-bipyramid if and only if its basis Q is an $(e - 1)$-polytope.

The facial structure of a bipyramid is determined by the facial structure of its basis as follows:

Theorem 7.8. *Let P be a bipyramid in $\mathbb{R}^d$ with basis Q and apices x_0 and x_1. Then the following holds:*

(a) *P is a polytope with* $\text{ext } P = (\text{ext } Q) \cup \{x_0, x_1\}$.
(b) *A subset F of P with $x_0, x_1 \notin F$ is a face of P if and only if it is a face of Q with $F \neq Q$.*
(c) *A subset F of P with $x_0 \in F$ and $x_1 \notin F$ is a face of P if and only if there is a face G of Q with $G \neq Q$ such that $F = \text{conv}(G \cup \{x_0\})$, i.e. $F = \{x_0\}$ or F is a pyramid with a face G of Q with $G \neq Q$ as the basis and x_0 as the apex. For each such face F of P, the face G is unique, and* $\dim G = \dim F - 1$. *Similarly for subsets F of P with $x_1 \in F$ and $x_0 \notin F$.*
(d) *A subset F of P with $x_0, x_1 \in F$ is a face of P if and only if $F = P$.*

PROOF. The proof follows the same lines as the proof of Theorem 7.7. The details are left to the reader. □

EXERCISES

7.1. Show that every polytope P with n vertices is the orthogonal projection of an $(n-1)$-simplex. (This is to be understood as follows: "Embed" P in $\mathbb{R}^{n-1}$; construct an $(n-1)$-simplex in $\mathbb{R}^{n-1}$ whose orthogonal projection onto aff P is P.)

7.2. Let $1 \leq n \leq d$. Starting with a $(d-n)$-polytope Q in $\mathbb{R}^d$, we may successively build up pyramids $P_1, P_2, \ldots, P_n$ by taking P_1 to be a $(d-n+1)$-pyramid with Q as a basis, taking P_2 to be a $(d-n+2)$-pyramid with P_1 as a basis, etc. The d-pyramid P_n is then called an *n-fold d-pyramid* with Q as a basis. Show that a $(d-1)$-fold d-pyramid is also a d-fold d-pyramid; it is, in fact, a d-simplex.

7.3. Copying Exercise 7.2, define the notion of a *n-fold d-bipyramid*. Show that a $(d-1)$-fold d-bipyramid is also a d-fold d-bipyramid.

(A particular type of d-fold d-bipyramids are the *d-crosspolytopes*; these are the convex hulls of $2d$ points $a_1, \ldots, a_d, b_1, \ldots, b_d$ such that all segments $[a_i, b_i]$ have a common midpoint, and no $[a_i, b_i]$ is contained in the affine hull of $[a_1, b_1], \ldots, [a_{i-1}, b_{i-1}]$. If the segments $[a_i, b_i]$ are orthogonal and have the same length, then the d-crosspolytope is said to be *regular*. A 3-crosspolytope is called an *octahedron*.)

7.4. A *prism* in $\mathbb{R}^d$ is a polytope of the form

$$P = \text{conv}(Q \cup (a + Q)),$$

where Q is a polytope in $\mathbb{R}^d$ with $\dim Q < d$, and $a + Q \not\subset \text{aff } Q$. Show that this definition is equivalent to the following: A prism in $\mathbb{R}^d$ is a polytope of the form

$$P = Q + [o, a],$$

where Q is a polytope in $\mathbb{R}^d$ with $\dim Q < d$ and a is a point in $\mathbb{R}^d \setminus \{o\}$ such that the line through o and a is not parallel to aff Q.

Show that

$$\text{ext } P = \text{ext } Q \cup \text{ext}(a + Q).$$

Show that the faces of P are the faces of Q, the faces of $a + Q$, and the prisms

$$F = \operatorname{conv}(G \cup (a + G)),$$

where G is a face of Q.

7.5. Copying Exercise 7.2, define the notion of an *n-fold d-prism*. Show that a $(d - 1)$-fold d-prism is a d-fold d-prism.

(A d-fold d-prism is also called a *d-parallellotope*; equivalently, a d-parallellotope is a polytope of the form

$$a + [o, b_1] + \cdots + [o, b_d],$$

where b_i is not contained in the affine hull of $a + [o, b_1] + \cdots + [o, b_{i-1}]$. If the segments $[o, b_i]$ are orthogonal and have the same length, then the d-parallellotope is called a *d-cube*.)

§8. Polyhedral Sets

In previous sections we have proved that every compact convex set C has an "external representation" as the intersection of closed halfspaces, namely, the supporting halfspaces, and an "internal representation" as the convex hull of a (unique) minimal set, namely, the set of extreme points. (Actually, for the external representation compactness is not needed, closedness suffices.) The sets which have a "finite" internal representation are the polytopes. In this section we shall study the sets which have a "finite" external representation, i.e. sets which are intersections of a finite number of closed halfspaces. These sets are called polyhedral sets. The main basic fact in polytope theory is that the polytopes are precisely the non-empty bounded polyhedral sets. Part of this statement will be proved at the end of this section; the remaining part will be proved in the next section.

A subset Q of $\mathbb{R}^d$ is called a *polyhedral set* if Q is the intersection of a finite number of closed halfspaces or $Q = \mathbb{R}^d$.

Every hyperplane H in $\mathbb{R}^d$ is the intersection of the two closed halfspaces which are bounded by H, and every affine subspace A of $\mathbb{R}^d$ with $A \neq \mathbb{R}^d$ is the intersection of a finite number of hyperplanes. Therefore, every affine subspace of $\mathbb{R}^d$ is polyhedral.

Let Q be a polyhedral set in $\mathbb{R}^d$, and let A be an affine subspace of $\mathbb{R}^d$ such that $Q \subset A \neq \mathbb{R}^d$. Then Q is the intersection of a finite number of closed halfspaces in A or $Q = A$. This follows from the fact that if K is a closed halfspace in $\mathbb{R}^d$ such that $A \cap K \neq \emptyset$, then $A \cap K$ is a closed halfspace in A or $A \cap K = A$.

Conversely, let A be an affine subspace of $\mathbb{R}^d$ with $A \neq \mathbb{R}^d$, and let Q be a subset of A such that Q is the intersection of a finite number of closed halfspaces in A or $Q = A$. Then Q is the intersection of a finite number of closed halfspaces in $\mathbb{R}^d$ and hence polyhedral. This follows from the fact that for every closed halfspace K in A there is a closed halfspace K' in $\mathbb{R}^d$ such that $A \cap K' = K$.

Every polyhedral set is closed and convex. The intersection of a finite number of polyhedral sets is again polyhedral. Any translate of a polyhedral set is again polyhedral. The image of a polyhedral set under an affine mapping is again polyhedral.

The facial structure of a (non-empty) polyhedral set Q in $\mathbb{R}^d$ is trivial when Q is an affine subspace of $\mathbb{R}^d$, the only faces being $\emptyset$ and Q. When Q is an e-dimensional polyhedral set in $\mathbb{R}^d$ which is not an affine subspace, then Q is affinely isomorphic to a polyhedral set Q' in $\mathbb{R}^e$ with $\dim Q' = e$ and $Q' \neq \mathbb{R}^e$. Therefore, when studying facial properties of polyhedral sets, it suffices to consider polyhedral sets Q in $\mathbb{R}^d$ with $\dim Q = d$ and $Q \neq \mathbb{R}^d$.

Every polyhedral set Q in $\mathbb{R}^d$ has a representation

$$Q = \bigcap_{i=1}^{n} K(x_i, \alpha_i). \tag{1}$$

In the following, when talking about a representation (1) of Q, we shall always implicitly assume that no two $K(x_i, \alpha_i)$'s are identical. For $Q \neq \mathbb{R}^d$ we may always assume that each $K(x_i, \alpha_i)$ is a closed halfspace, i.e. each x_i is $\neq o$. For $Q = \mathbb{R}^d$ there is only one representation, namely, $Q = K(o, \alpha)$, where $\alpha \geq 0$. Note that when $Q \neq \mathbb{R}^d$ there are infinitely many representations (unless $d = 0$); new closed halfspaces containing Q may always be added.

We shall call a representation (1) *irreducible* if $n = 1$, or $n > 1$ and

$$Q \subsetneqq \bigcap_{\substack{i=1 \\ i \neq j}}^{n} K(x_i, \alpha_i), \qquad j = 1, \ldots, n.$$

A representation which is not irreducible is called *reducible*. Clearly, any reducible representation may be turned into an irreducible representation by omitting certain of the sets $K(x_i, \alpha_i)$. It follows from Theorem 8.2 below that there is only one irreducible representation of each polyhedral set Q which is not an affine subspace.

Theorem 8.1. *Let Q be a polyhedral set in $\mathbb{R}^d$ with* $\dim Q = d$ *and $Q \neq \mathbb{R}^d$. Let*

$$Q = \bigcap_{i=1}^{n} K(x_i, \alpha_i)$$

be a representation of Q with $n > 1$, where each $K(x_i, \alpha_i)$ is a closed halfspace. Then the representation is irreducible if and only if

$$H(x_j, \alpha_j) \cap \operatorname{int} \bigcap_{\substack{i=1 \\ i \neq j}}^{n} K(x_i, \alpha_i) \neq \emptyset$$

for each $j = 1, \ldots, n$.

Proof. For $j = 1, \ldots, n$, we let

$$M_j := \bigcap_{\substack{i=1 \\ i \neq j}}^{n} K(x_i, \alpha_i).$$

Then $Q = K(x_j, \alpha_j) \cap M_j$ for each j, and since $\dim Q = d$ by assumption, we see that $\operatorname{int} M_j \neq \emptyset$; consequently, $\operatorname{ri} M_j = \operatorname{int} M_j$ and $M_j \not\subset H(x_j, \alpha_j)$.

The condition of the theorem reads

$$H(x_j, \alpha_j) \cap \operatorname{int} M_j \neq \emptyset, \qquad j = 1, \ldots, n. \tag{2}$$

By Theorem 4.1 and the observations above, (2) is equivalent to

$$M_j \not\subset K(x_j, \alpha_j), \qquad M_j \not\subset K(-x_j, -\alpha_j), \qquad j = 1, \ldots, n. \tag{3}$$

Now, $M_j \subset K(-x_j, -\alpha_j)$ would imply

$$Q \subset K(x_j, \alpha_j) \cap K(-x_j, -\alpha_j) = H(x_j, \alpha_j),$$

a contradiction. Hence, (3) is equivalent to

$$M_j \not\subset K(x_j, \alpha_j), \qquad j = 1, \ldots, n. \tag{4}$$

But (4) is just a rephrasing of irreducibility. □

The following theorem shows that the boundary of a polyhedral set is built up in the expected way:

Theorem 8.2. *Let Q be a polyhedral set in $\mathbb{R}^d$ with* $\dim Q = d$ *and $Q \neq \mathbb{R}^d$. Let*

$$Q = \bigcap_{i=1}^{n} K(x_i, \alpha_i) \tag{*}$$

be a representation of Q, where each $K(x_i, \alpha_i)$ is a closed halfspace. Then the following holds:

(a) $\operatorname{bd} Q = \bigcup_{i=1}^{n} H(x_i, \alpha_i) \cap Q$.
(b) *Each facet of Q is of the form $H(x_j, \alpha_j) \cap Q$.*
(c) *Each set $H(x_j, \alpha_j) \cap Q$ is a facet of Q if and only if the representation (*) is irreducible.*

Proof. (a) We have

$$\begin{aligned} \operatorname{int} Q &= \operatorname{int} \bigcap_{i=1}^{n} K(x_i, \alpha_i) \\ &= \bigcap_{i=1}^{n} \operatorname{int} K(x_i, \alpha_i) \\ &= \bigcap_{i=1}^{n} K(x_i, \alpha_i) \setminus H(x_i, \alpha_i) \end{aligned}$$

which implies (a).

(b) Let F be a facet of Q. Let x be a relative interior point of F. Then F is the smallest face of Q containing x, cf. Theorem 5.6. By (a), there is j such that

$$x \in H(x_j, \alpha_j) \cap Q.$$

But then we must have

$$F \subset H(x_j, \alpha_j) \cap Q,$$

implying

$$F = H(x_j, \alpha_j) \cap Q,$$

cf. Corollary 5.5.

(c) For $n = 1$ there is nothing to prove. So assume that $n > 1$.

If $(*)$ is irreducible, then each $H(x_j, \alpha_j)$ supports Q, whence $H(x_j, \alpha_j) \cap Q$ is a proper face of Q. We prove that $H(x_j, \alpha_j) \cap Q$ has a non-empty interior in $H(x_j, \alpha_j)$; this will imply that $H(x_j, \alpha_j) \cap Q$ is a facet. We have

$$\begin{aligned} H(x_j, \alpha_j) \cap Q &= H(x_j, \alpha_j) \cap \bigcap_{i=1}^{n} K(x_i, \alpha_i) \\ &= H(x_j, \alpha_j) \cap \bigcap_{\substack{i=1 \\ i \neq j}}^{n} K(x_i, \alpha_i) \\ &\supset H(x_j, \alpha_j) \cap \operatorname{int} \bigcap_{\substack{i=1 \\ i \neq j}}^{n} K(x_i, \alpha_i) \\ &\neq \emptyset, \end{aligned}$$

cf. Theorem 8.1. Since the set

$$H(x_j, \alpha_j) \cap \operatorname{int} \bigcap_{\substack{i=1 \\ i \neq j}}^{n} K(x_i, \alpha_i)$$

is open in $H(x_j, \alpha_j)$, the desired conclusion follows.

Conversely, if $(*)$ is reducible, then

$$Q = \bigcap_{\substack{i=1 \\ i \neq j}}^{n} K(x_i, \alpha_i)$$

for some j. Suppose that $H(x_j, \alpha_j) \cap Q$ is a facet of Q. Let x be a relative interior point of $H(x_j, \alpha_j) \cap Q$. Using (a) we see that there is an i with $i \neq j$ and $x \in H(x_i, \alpha_i) \cap Q$. But then we must have

$$H(x_j, \alpha_j) \cap Q = H(x_i, \alpha_i) \cap Q,$$

cf. Corollary 5.5. This, however, implies

$$K(x_j, \alpha_j) = K(x_i, \alpha_i),$$

a contradiction. Hence, $H(x_j, \alpha_j) \cap Q$ is not a facet of Q. □

The preceding theorem shows that most polyhedral sets have facets, the only exceptions being affine subspaces.

Theorem 8.3. *Let F be a proper face of a polyhedral set Q in $\mathbb{R}^d$. Then there is a facet G of Q containing F.*

PROOF. We may assume that $\dim Q = d$. Choose an irreducible representation

$$Q = \bigcap_{i=1}^{n} K(x_i, \alpha_i).$$

Let x be a relative interior point of F. By Theorem 8.2(a), there is j such that

$$x \in H(x_j, \alpha_j) \cap Q.$$

Now, F is the smallest face containing x, cf. Theorem 5.6, and $H(x_j, \alpha_j) \cap Q$ is a facet containing x, cf. Theorem 8.2(c). Therefore, with

$$G := H(x_j, \alpha_j) \cap Q$$

we have the desired conclusion. □

Corollary 8.4. *Let Q be a polyhedral set in $\mathbb{R}^d$. Then every face of Q is also a polyhedral set.*

PROOF. We need only prove the statement for proper faces of Q. Theorem 8.3 shows that any proper face of Q is a face of a facet of Q. Facets of Q, however, are polyhedral sets by Theorem 8.2(b). The statement then follows by induction on the dimension. □

Corollary 8.5. *Let Q be a polyhedral set in $\mathbb{R}^d$. Then the number of faces of Q is finite.*

PROOF. The number of facets of a polyhedral set Q is finite, cf. Theorem 8.2(b). Each proper face of Q is a face of a facet of Q by Theorem 8.3. The statement then follows by induction on the dimension. □

Corollary 8.6. *Let Q be a polyhedral set in $\mathbb{R}^d$ with* $\dim Q = d$. *Let F_j and F_k be faces of Q with*

$$F_j \subset F_k$$

and

$$\dim F_j = j, \qquad \dim F_k = k,$$

where

$$0 \leq j < j + 1 \leq k - 1 < k \leq d.$$

Then there are faces $F_{j+1}, \ldots, F_{k-1}$ of Q with

$$F_j \subset F_{j+1} \subset \cdots \subset F_{k-1} \subset F_k$$

and

$$\dim F_i = i, \qquad i = j + 1, \ldots, k - 1.$$

Proof. By Theorem 5.2, F_j is a proper face of F_k. And by Corollary 8.4, F_k is polyhedral. Theorem 8.3 then ensures the existence of a facet F_{k-1} of F_k with $F_j \subset F_{k-1}$. If $j = k - 2$, we have the desired conclusion. If $j < k - 2$, we argue as above with F_{k-1} replacing F_k. Continuing this way, we obtain faces F_i with the desired properties. □

In Corollary 8.6, note that we actually have

$$F_j \subsetneq F_{j+1} \subsetneq \cdots \subsetneq F_{k-1} \subsetneq F_k.$$

Note also that the statement is not valid in general with $j = -1$.

We conclude this section with the following:

Corollary 8.7. *Let Q be a non-empty bounded polyhedral set in $\mathbb{R}^d$. Then Q is a polytope.*

Proof. By assumption, Q is a compact convex set. By Corollary 8.5, ext Q is a finite set. The statement then follows from Theorem 7.1, (b) $\Rightarrow$ (a). □

The converse of Corollary 8.7 is also valid, see Section 9.

Exercises

8.1. Show that the image of a polyhedral set under an affine mapping is again a polyhedral set.

8.2. Show that every face of a non-empty polyhedral set is exposed.

8.3. Show that every non-empty polyhedral set not containing any line has at least one vertex. (Here, of course, a vertex of a polyhedral set means a 0-dimensional face, exposed by Exercise 8.2.)

§9. Polarity of Polytopes and Polyhedral Sets

In this section we shall apply the polarity theory of Section 6 to polytopes and polyhedral sets. We shall show that the polar of a polytope with o as an interior point is a bounded polyhedral set with o as an interior point, and conversely. As promised in Section 8, we shall deduce that every polytope is a bounded polyhedral set (whence polytopes can also be described as non-empty bounded polyhedral sets). Furthermore, we shall improve a result of Section 6 by showing that $\dim F + \dim G = d - 1$ when F and G are conjugate faces of mutually polar d-polytopes.

The following theorem explains in detail the polarity of convex polytopes and polyhedral sets. Note that the polyhedral sets Q having a representation of the particular form

$$Q = \bigcap_{i=1}^{n} K(x_i, 1)$$

are precisely the polyhedral sets which have o as an interior point.

Theorem 9.1. *Let $x_1, \ldots, x_n$, where $n \geq 1$, be distinct points of $\mathbb{R}^d$, and let*

$$P := \operatorname{conv}\{x_1, \ldots, x_n\},$$

$$Q := \bigcap_{i=1}^{n} K(x_i, 1).$$

Then we have:

(a) $P^\circ = Q$.
(b) $Q^\circ = \operatorname{conv}\{o, x_1, \ldots, x_n\}$.
(c) *P and Q are mutually polar sets if and only if $o \in P$.*
(d) *P and Q are mutually polar sets with Q bounded if and only if $o \in \operatorname{int} P$.*
(e) *Suppose that P and Q are mutually polar sets with Q bounded* (i.e. $o \in \operatorname{int} P$, *cf.* (d)). *Then we have*

$$\operatorname{ext} P = \{x_1, \ldots, x_n\}$$

if and only if the representation

$$Q = \bigcap_{i=1}^{n} K(x_i, 1)$$

is irreducible.

Proof. (a) Formula (1) of Section 6 shows that

$$\{x_1, \ldots, x_n\}^\circ = Q, \tag{1}$$

and formula (2) of Section 6 shows that

$$\{x_1, \ldots, x_n\}^\circ = (\operatorname{conv}\{x_1, \ldots, x_n\})^\circ \tag{2}$$

since $M \subset K(y, 1)$ if and only if $\operatorname{conv} M \subset K(y, 1)$. Combining (1) and (2) we obtain (a).

(b) Using (a), Theorem 6.2 and Corollary 2.9 we have

$$\begin{aligned} Q^\circ &= P^{\circ\circ} \\ &= \operatorname{clconv}\{o, x_1, \ldots, x_n\} \\ &= \operatorname{conv}\{o, x_1, \ldots, x_n\} \end{aligned}$$

which proves (b).

(c) This is an immediate consequence of (a) and (b).

(d) This follows from (c) and Theorem 6.1.

(e) By assumption we must have $n \geq 2$. For $j = 1, \ldots, n$, let

$$P_j := \operatorname{conv}\{x_1, \ldots, x_{j-1}, x_{j+1}, \ldots, x_n\},$$

$$Q_j := \bigcap_{\substack{i=1 \\ i \neq j}}^{n} K(x_i, 1).$$

Note that application of (a) to $\{x_1, \ldots, x_{j-1}, x_{j+1}, \ldots, x_n\}$ instead of $\{x_1, \ldots, x_n\}$ gives

$$P_j^\circ = Q_j. \tag{3}$$

Furthermore, Theorem 7.2, (a) $\Rightarrow$ (b) shows that we always have

$$\operatorname{ext} P \subset \{x_1, \ldots, x_n\}. \tag{4}$$

Now, if $\operatorname{ext} P$ is a proper subset of $\{x_1, \ldots, x_n\}$, then $P = P_j$ for some j by Theorem 7.2(c). Then also $P^\circ = P_j^\circ$, whence $P^\circ = Q_j$ by (3). But $P^\circ = Q$ by (a), and therefore we have $Q = Q_j$. This shows that the representation of Q is reducible.

Conversely, if the representation of Q is reducible, then $Q = Q_j$ for some j, and so also $Q^\circ = Q_j^\circ$. Application of (b) to $\{x_1, \ldots, x_{j-1}, x_{j+1}, \ldots, x_n\}$ instead of $\{x_1, \ldots, x_n\}$ gives

$$Q_j^\circ = \operatorname{conv}\{o, x_1, \ldots, x_{j-1}, x_{j+1}, \ldots, x_n\}.$$

Since $Q^\circ = P$ by assumption, and $Q^\circ = Q_j^\circ$, as we just have seen, it follows that

$$P = \operatorname{conv}\{o, x_1, \ldots, x_{j-1}, x_{j+1}, \ldots, x_n\}.$$

Now, Theorem 7.2, (a) $\Leftrightarrow$ (b) shows that here any non-extreme point of P among the points $o, x_1, \ldots, x_{j-1}, x_{j+1}, \ldots, x_n$ may be omitted. It follows from (4) and the assumption that o is such a point. Therefore,

$$P = \operatorname{conv}\{x_1, \ldots, x_{j-1}, x_{j+1}, \ldots, x_n\}.$$

Theorem 7.2, (a) $\Rightarrow$ (b) then shows that

$$\operatorname{ext} P \subset \{x_1, \ldots, x_{j-1}, x_{j+1}, \ldots, x_n\},$$

whence $\operatorname{ext} P$ is a proper subset of $\{x_1, \ldots, x_n\}$. □

We are now ready to prove the following main theorem:

Theorem 9.2. *A non-empty subset P of $\mathbb{R}^d$ is a polytope if and only if it is a bounded polyhedral set.*

Proof. We have already proved the "if" statement in Corollary 8.7. Conversely, let P be a polytope in $\mathbb{R}^d$, say

$$P = \operatorname{conv}\{x_1, \ldots, x_n\}.$$

To prove that P is a bounded polyhedral set it causes no loss of generality to assume that $o \in \operatorname{int} P$. Theorem 9.1(a) shows that $P^\circ = Q$, where Q denotes the polyhedral set defined by

$$Q := \bigcap_{i=1}^{n} K(x_i, 1).$$

It follows from Theorem 9.1(d) that Q is bounded and that $Q^\circ = P$. Applying now Corollary 8.7 to Q, it follows that Q is a polytope, say

$$Q = \operatorname{conv}\{y_1, \ldots, y_m\}.$$

We next apply Theorem 9.1 to $\{y_1, \ldots, y_m\}$ instead of $\{x_1, \ldots, x_n\}$. Statement (a) shows that $Q^\circ = R$, where R denotes the polyhedral set defined by

$$R := \bigcap_{j=1}^{m} K(y_j, 1).$$

But we have already seen that $Q^\circ = P$, whence

$$P = \bigcap_{j=1}^{m} K(y_j, 1),$$

i.e. P is a polyhedral set. □

We may now use the results of Section 8 on polyhedral sets to obtain results on polytopes.

Corollary 9.3. *Let P_1 and P_2 be polytopes in $\mathbb{R}^d$ such that $P_1 \cap P_2 \neq \emptyset$. Then $P_1 \cap P_2$ is also a polytope.*

PROOF. The intersection of any two polyhedral sets in $\mathbb{R}^d$ is polyhedral. The statement then follows from Theorem 9.2. □

Corollary 9.4. *Let P be a polytope in $\mathbb{R}^d$, and let A be an affine subspace of $\mathbb{R}^d$ such that $P \cap A \neq \emptyset$. Then $P \cap A$ is also a polytope.*

PROOF. Any affine subspace A of $\mathbb{R}^d$ is polyhedral. The statement then follows as in the proof of Corollary 9.3. □

Corollary 9.5. *Let P be a d-polytope in $\mathbb{R}^d$. Then P has at least $d + 1$ facets.*

PROOF. Let

$$P = \operatorname{conv}\{x_1, \ldots, x_n\},$$

and assume without loss of generality that $o \in \operatorname{int} P$. Let

$$Q := \bigcap_{i=1}^{n} K(x_i, 1).$$

Then by Theorem 9.1(d), P and Q are mutually polar sets with Q bounded (and $o \in \operatorname{int} Q$). Corollary 8.7 next shows that Q is a d-polytope. Let $y_1, \ldots, y_m$ be the vertices of Q, and let

$$R := \bigcap_{j=1}^{m} K(y_j, 1).$$

Then Q and R are mutually polar by Theorem 9.1(d), whence $R = P$. Moreover, Theorem 9.1(e) shows that the representation

$$P = \bigcap_{j=1}^{m} K(y_j, 1),$$

is irreducible. But then the number of facets of P is m by Theorem 8.2(b), (c). On the other hand, the number of vertices of the d-polytope Q is at least $d + 1$, whence $m \geq d + 1$, as desired. □

In the next corollary, note that when F is a facet of a d-polytope in $\mathbb{R}^d$, then aff F is a supporting hyperplane of P.

Corollary 9.6. *Let P be a d-polytope in $\mathbb{R}^d$, let $F_1, \ldots, F_n$ be the facets of P, and let $K(x_i, \alpha_i)$ be the supporting halfspace of P bounded by* aff *F_i for $i = 1, \ldots, n$. Then*

$$P = \bigcap_{i=1}^{n} K(x_i, \alpha_i),$$

and this representation is irreducible.

PROOF. By Theorem 9.2, P is polyhedral. Let

$$P = \bigcap_{j=1}^{m} K(y_j, \beta_j)$$

be an irreducible representation of P. By Theorem 8.2(b), (c), the facets of P are the sets $H(y_j, \beta_j) \cap P$. But the facets of P are also the sets $H(x_i, \alpha_i) \cap P$ by assumption. Therefore, $m = n$ and there is a one-to-one correspondence between the i's and j's such that

$$H(x_i, \alpha_i) = H(y_j, \beta_j)$$

for corresponding i and j. Then of course also

$$K(x_i, \alpha_i) = K(y_j, \beta_j)$$

for corresponding i and j. This shows that

$$P = \bigcap_{i=1}^{n} K(x_i, \alpha_i),$$

and that this representation is irreducible. □

Corollary 9.7. *Let P be a d-polytope in $\mathbb{R}^d$. Let F_j and F_k be faces of P with*

$$F_j \subset F_k$$

and

$$\dim F_j = j, \qquad \dim F_k = k,$$

where

$$-1 \leq j < j + 1 \leq k - 1 < k \leq d.$$

Then there are faces $F_{j+1}, \ldots, F_{k-1}$ of P with

$$F_j \subset F_{j+1} \subset \cdots \subset F_{k-1} \subset F_k$$

and

$$\dim F_i = i, \qquad i = j + 1, \ldots, k - 1.$$

Proof. With Theorem 9.2 in mind, the statement follows immediately from Corollary 8.6 when $j \geq 0$. For $j = -1$, let F_0 be any vertex of F_k. If $k = 1$, we have the desired conclusion. If $k \geq 2$, apply Corollary 8.6 to the faces F_0 and F_k. □

We shall finally improve the inequality of Theorem 6.10. Note that when P is a d-polytope in $\mathbb{R}^d$ with $o \in \operatorname{int} P$, then P and P° form a pair of mutually polar d-polytopes, and each pair of mutually polar d-polytopes arises in this way; this follows from Theorem 9.1(a), (d) and Theorem 9.2. We also remind the reader that any face F of P is a member of a pair F, G of conjugate faces since all faces of P are exposed, cf. Theorem 7.5.

Theorem 9.8. *Let P and Q be mutually polar d-polytopes in $\mathbb{R}^d$ and let F and G be conjugate faces of P and Q, respectively. Then*

$$\dim F + \dim G = d - 1.$$

In particular, vertices of P are conjugate to facets of Q, and facets of P are conjugate to vertices of Q.

Proof. We shall appeal to the proof of Theorem 6.10. As explained there, we need only consider the case where F and G are proper faces. Let $x_1, \ldots, x_n$ be the vertices of P, and let $x_1, \ldots, x_k$ be the vertices of F. Then by Theorem 9.1(a),

$$Q = \bigcap_{i=1}^{n} K(x_i, 1). \tag{5}$$

Moreover, by the definition of the $\triangle$-operation, we have

$$G = Q \cap \bigcap_{x \in F} H(x, 1). \tag{6}$$

Now, note that

$$\bigcap_{x \in F} H(x, 1) = \bigcap_{i=1}^{k} H(x_i, 1). \tag{7}$$

In fact, if y is in $H(x_i, 1)$ for $i = 1, \ldots, k$, then $x_1, \ldots, x_k$ are in $H(y, 1)$, whence every $x \in F$ must be in $H(y, 1)$, showing that y is in $H(x, 1)$ for every $x \in F$. Combining now (5), (6) and (7) we obtain

$$G = \bigcap_{i=k+1}^{n} K(x_i, 1) \cap \bigcap_{i=1}^{k} H(x_i, 1). \tag{8}$$

Let

$$A := \bigcap_{i=1}^{k} H(x_i, 1).$$

Then A is an affine subspace containing G. In fact, $A = \text{aff}\, G$. To see this, note first that (8) shows that G is a polyhedral set in A with the representation

$$G = \bigcap_{i=k+1}^{n} K(x_i, 1) \cap A$$

(where, of course, we may have $K(x_i, 1) \cap A = A$ for certain values of i). Now, it is clear that the non-empty intersection G of closed halfspaces in the affine space A can only have a smaller dimension than A itself if G is contained in a hyperplane bounding one of the halfspaces $K(x_i, 1) \cap A$, $i = k + 1, \ldots, n$. But if G is a subset of $H(x_i, 1)$, then $x_i \in G^{\Delta} = F$, a contradiction. In conclusion, $A = \text{aff}\, G$, whence in particular

$$\begin{aligned} \dim G &= \dim A \\ &= \dim \bigcap_{x \in F} H(x, 1). \end{aligned}$$

Hence, in the proof of Theorem 6.10 we have equality in (9). The rest of the proof of Theorem 6.10 then yields the desired formula. □

For another proof of Theorem 9.8, see Section 10.

Note that the proof of Corollary 9.5 could also have been based on Theorem 9.8.

Exercises

9.1. A *section* of a polytope P in $\mathbb{R}^d$ is the intersection of P and some affine subspace of $\mathbb{R}^d$. Show that every polytope P with n facets is a section of an $(n-1)$-simplex. (This is to be understood as follows: "Embed" P in $\mathbb{R}^{n-1}$; construct an $(n-1)$-simplex S in $\mathbb{R}^{n-1}$ such that $S \cap \text{aff}\, P = P$. Hint: One may use Exercises 6.5 and 7.1.)

9.2. Let P and Q be mutually polar convex polytopes in $\mathbb{R}^d$, and let F and G be conjugate faces of P and Q, respectively. Show that $G = Q \cap H(x, 1)$ if and only if $x \in \text{ri}\, F$, cf. Exercise 6.6.

§10. Equivalence and Duality of Polytopes

It may be said that the combinatorial theory of convex polytopes (which is the subject of Chapter 3) is the study of face-lattices of convex polytopes. So, from this point of view, there is no need to distinguish between polytopes whose face-lattices are isomorphic. This leads to the notion of equivalent polytopes.

In Section 6 we developed a polarity theory of convex sets, and in Section 9 we applied it to obtain basic properties of convex polytopes. It is a fundamental fact that for a pair P, Q of mutually polar polytopes, the $\triangle$-operation induces an anti-isomorphism of the face-lattices. Accepting the point of view explained above (leading to the notion of equivalence), it follows that there is no need to distinguish between Q and any polytope whose face-lattice is isomorphic to that of Q. These polytopes, however, are just the polytopes whose face-lattices are anti-isomorphic to that of P. This leads to the notion of dual polytopes.

Two polytopes are said to be *equivalent* (and each is said to be an equivalent of the other) if their face-lattices are isomorphic. Clearly, this is an equivalence relation. The image $\varphi(P)$ of a polytope P under an affine isomorphism φ is an equivalent of P; but in general there are many other equivalents of P.

Theorem 10.1. *Let P and Q be equivalent polytopes with* $\dim P = d$, *and let*

$$\varphi: (\mathscr{F}(P), \subset) \to (\mathscr{F}(Q), \subset)$$

be an isomorphism. Then

$$\dim Q = d,$$

and

$$\dim \varphi(F) = \dim F$$

for any face F of P.

PROOF. By Corollary 9.7, each face F of P is a member of a chain

$$\varnothing = F_{-1} \subsetneqq \cdots \subsetneqq F_i \subsetneqq \cdots \subsetneqq F_d = P \tag{1}$$

of faces of P with

$$\dim F_i = i, \qquad i = -1, \ldots, d.$$

Since φ is an isomorphism, (1) yields

$$\varnothing = \varphi(F_{-1}) \subsetneqq \cdots \subsetneqq \varphi(F_i) \subsetneqq \cdots \subsetneqq \varphi(F_d) = Q. \tag{2}$$

Now, (2) implies

$$\dim \varphi(F_{i+1}) \geq \dim \varphi(F_i) + 1, \qquad i = -1, \ldots, d-1, \tag{3}$$

cf. Corollary 5.5. This clearly implies $\dim \varphi(F_d) \geq d$, i.e.

$$\dim Q \geq d = \dim P.$$

But P and Q play completely symmetric roles, and therefore we must actually have

$$\dim Q = d,$$

as desired. This, in turn, implies that we must have equality in (3) for all i, enforcing

$$\dim \varphi(F_i) = i, \qquad i = -1, \ldots, d.$$

This completes the proof. □

Two polytopes are said to be *dual* (and each is said to be a dual of the other) if their face-lattices are anti-isomorphic. We note that when P and Q_1 are dual, then P and Q_2 are also dual if and only if Q_1 and Q_2 are equivalent.

The question of existence has almost been settled by Corollary 6.8:

Theorem 10.2. *For any polytope P, there is a dual polytope Q.*

PROOF. For any d-polytope P there is a d-polytope P_1 in $\mathbb{R}^d$ with $o \in \operatorname{int} P_1$ such that P and P_1 are equivalent. Corollary 6.8 shows that $Q_1 := P_1^\circ$ is a dual of P_1. But then Q_1 is also a dual of P by the equivalence of P and P_1. □

The next theorem is closely related to Theorem 9.8, see the remarks below:

Theorem 10.3. *Let P and Q be dual polytopes with* $\dim P = d$, *and let*

$$\psi\colon (\mathscr{F}(P), \subset) \to (\mathscr{F}(Q), \subset)$$

be an anti-isomorphism. Then

$$\dim Q = d,$$

and

$$\dim \psi(F) = d - 1 - \dim F$$

for any face F of P.

PROOF. As in the proof of Theorem 10.1, we use the fact that each face F of P is a member of a chain

$$\varnothing = F_{-1} \subsetneqq \cdots \subsetneqq F_i \subsetneqq \cdots \subsetneqq F_d = P \tag{4}$$

of faces of P with

$$\dim F_i = i, \qquad i = -1, \ldots, d. \tag{5}$$

Since ψ is an anti-isomorphism, (4) yields

$$Q = \psi(F_{-1}) \supsetneqq \cdots \supsetneqq \psi(F_i) \supsetneqq \cdots \supsetneqq \psi(F_d) = \varnothing. \tag{6}$$

Now, (6) implies

$$\dim \psi(F_i) \geq \dim \psi(F_{i+1}) + 1, \qquad i = -1, \ldots, d-1. \tag{7}$$

This clearly implies $\dim \psi(F_{-1}) \geq d$, i.e.

$$\dim Q \geq d = \dim P.$$

The symmetry of P and Q then ensures that we must have

$$\dim Q = d,$$

proving the first statement. This, in turn, implies that we must have equality in (7) for all i, whence

$$\begin{aligned}\dim \psi(F_i) &= d - 1 - i \\ &= d - 1 - \dim F_i, \qquad i = -1, \ldots, d,\end{aligned}$$

completing the proof. □

It is clear that Theorem 9.8 is an immediate consequence of Theorem 10.3. On the other hand, Theorem 10.3 could also have been deduced from Theorem 10.1; just observe that when P and Q are dual with $\dim P = d$, then there is a pair P_1, Q_1 of mutually polar d-polytopes such that P is equivalent to P_1 and Q is equivalent to Q_1.

We next prove two important theorems on the facial structure of polytopes; their proofs depend on Theorem 10.3.

Theorem 10.4. *Let P be a d-polytope, and let F be a proper face of P. Then F is the intersection of the facets of P containing F. If F is a k-face, then for $k = 0, 1, \ldots, d-3$ there are at least $d-k$ such facets, for $k = d-2$ there are exactly 2 $(= d-k)$ such facets, and for $k = d-1$ there is exactly 1 $(= d-k)$ such facet.*

PROOF. Let Q be a dual of P, and let ψ be an anti-isomorphism from $(\mathscr{F}(P), \subset)$ onto $(\mathscr{F}(Q), \subset)$. Let F be a k-face of P, and let $G := \psi(F)$. Then

$$\dim G = d - 1 - k$$

by Theorem 10.3. But then G has at least

$$(d - 1 - k) + 1 = d - k$$

vertices. For $k = d-2$ and $k = d-1$, the number of vertices is exactly $d-k$, since 1-polytopes have two vertices and 0-polytopes have one vertex. We now apply the anti-isomorphism ψ^{-1} from $(\mathscr{F}(Q), \subset)$ onto $(\mathscr{F}(P), \subset)$. The dimension formula of Theorem 10.3 shows that vertices of Q correspond to facets of P. Therefore, since G is the smallest face of Q containing the vertices of G, it follows that F is the intersection of the facets of P containing F. This proves the statement. □

Theorem 10.5. *Let P be a d-polytope, and let x be a vertex of P. Then there are at least d edges of P containing x.*

PROOF. Let Q be a dual of P, and let ψ be an anti-isomorphism from $(\mathscr{F}(P), \subset)$ onto $(\mathscr{F}(Q), \subset)$. Let $G := \psi(\{x\})$; then by Theorem 10.3, G is a $(d-1)$-face of Q. Corollary 9.5 next shows that the number of $(d-2)$-faces of G is at least d. But then by duality the number of 1-faces of P containing x is at least d, cf. Theorem 10.3. □

Using terminology from graph theory (cf. Appendix 2 and Section 15), we shall say that two distinct vertices of a polytope P are *adjacent* if the segment joining them is an edge of P, and we shall say that a vertex and an edge are *incident* if the vertex is a vertex of the edge. With this terminology, Theorem 10.5 states that the number of vertices adjacent to x, or the number of edges incident to x, is at least d.

For any d-polytope P, let $f_j(P)$ denote the number of j-faces of P, where $j = -1, 0, \ldots$. Thus $f_{-1}(P) = f_d(P) = 1$ and $f_j(P) = 0$ for $j > d$. For $d \geq 1$, the d-tuple

$$f(P) := (f_0(P), \ldots, f_{d-1}(P))$$

of positive integers is called the *f-vector* of P. This concept will play a central role in Chapter 3. Here we note an immediate corollary of Theorems 10.1 and 10.3:

Corollary 10.6. *Let P be a d-polytope (where $d \geq 1$). Then for any polytope P_1 equivalent to P we have*

$$f(P_1) = (f_0(P), \ldots, f_{d-1}(P)),$$

i.e. $f(P_1) = f(P)$, and for any polytope Q dual to P we have

$$f(Q) = (f_{d-1}(P), \ldots, f_0(P)).$$

EXERCISES

10.1. Show that an equivalent of a d-pyramid is again a d-pyramid.

10.2. Show that a dual of a d-pyramid is again a d-pyramid.

10.3. Show that if Q_1 and Q_2 are equivalent polytopes, then any pyramid P_1 with Q_1 as a basis is equivalent to any pyramid P_2 with Q_2 as a basis.

10.4. Show that the statement of Exercise 10.1 is not valid in general for bipyramids.

10.5. Show that any bipyramid has prisms as well as non-prisms as duals.

10.6. Let P_1 and P_2 be polytopes, let φ' be a one-to-one mapping from the vertices of P_1 onto the vertices of P_2, and let φ'' be a one-to-one mapping from the facets of P_1 onto the facets of P_2. Assume that $\varphi'(x)$ is a vertex of $\varphi''(F)$ if and only if x is a vertex of F. Show that there is an isomorphism φ from $(\mathscr{F}(P_1), \subset)$ onto $(\mathscr{F}(P_2), \subset)$ which extends both φ' and φ''. In particular, P_1 and P_2 are equivalent. State and prove an analogous dual statement.

10.7. Use a duality argument to show that no vertex of a polytope P is contained in all the facets of P.

10.8. Let F_1 and F_2 be faces of a d-polytope P such that $F_1 \subset F_2$ and $\dim F_1 < \dim F_2 \leq d - 1$. Show that there is a face F_3 of P such that $\dim F_3 = \dim F_1 + 1$, $F_1 \subset F_3$ and $F_3 \not\subset F_2$. (Hint: Exercise 10.7 may be useful.)

10.9. Let P be a d-polytope. Show that for $j < k \leq d - 1$, any j-face F of P is the intersection of the k-faces of P containing F.

10.10. Let P be an octahedron and let Q be a 3-polytope obtained by "adding pyramids" over two of the facets of a 3-simplex. Show that $f(P) = f(Q)$. Show that P and Q are non-equivalent.

§11. Vertex-Figures

The vertex-figures of a d-polytope P are certain $(d - 1)$-polytopes, each containing information about the "local" facial structure of P "near" one of its vertices. In this section we have collected some results dealing with or related to vertex-figures.

We first study the facial structure of a non-empty intersection $H \cap P$ of a d-polytope P in $\mathbb{R}^d$ and a hyperplane H in $\mathbb{R}^d$. Note that when H does not intersect int P, then H must be a supporting hyperplane by Theorem 4.1, and $H \cap P$ is then simply a face of P (whose facial structure is known when the facial structure of P is known).

Theorem 11.1. *Let P be a d-polytope in $\mathbb{R}^d$, and let H be a hyperplane in $\mathbb{R}^d$ such that*

$$H \cap \operatorname{int} P \neq \emptyset.$$

Then the following holds:

(a) *The set $P' := H \cap P$ is a $(d - 1)$-polytope.*
(b) *Let F be a face of P. Then $F' := H \cap F$ is a face of P', and* $\dim F' \leq \dim F$. *If $F \neq \emptyset$ and H is not a supporting hyperplane of F (i.e. F' is not a face of F and hence not a face of P), then* $\dim F' = \dim F - 1$.
(c) *Let F' be a face of P'. Then there is at least one face F of P such that $F' = H \cap F$, and for each such face F we have* $\dim F \geq \dim F'$.
(d) *Let F' be a face of P'. If F' is not a face of P, then there is one and only one face F of P such that $F' = H \cap F$, and for this face F we have* $\dim F = \dim F' + 1$.

Proof. (a) The set P' is a polytope by Corollary 9.4. It is clear that the dimension of P' is $d - 1$.

(b) It follows immediately from the definition of a face that F' is a face of P', and it is clear that $\dim F' \le \dim F$. If $F \neq \emptyset$ and H is not a supporting hyperplane of F, then we have $F \not\subset H$ and $H \cap \operatorname{ri} F \neq \emptyset$, cf. Theorem 4.1. But then $H \cap \operatorname{aff} F$ is a hyperplane in $\operatorname{aff} F$ intersecting the interior of F in $\operatorname{aff} F$, whence $\dim(H \cap F) = \dim F - 1$.

(c) For $F' = \emptyset$, the statement is obvious. For $F' \neq \emptyset$, we first note that it is trivial that $\dim F \ge \dim F'$ when F is a face of P such that $F' = H \cap F$. To prove the existence of such a face, let $x_0 \in \operatorname{ri} F'$ and let F_0 be the smallest face of P containing x_0. Then $x_0 \in \operatorname{ri} F_0$, cf. Theorem 5.6. We shall complete the proof by showing that

$$F' = H \cap F_0. \tag{1}$$

Let $y \in F'$ with $y \neq x_0$. Then there exists a point $z \in F'$ such that $x_0 \in]y, z[$, cf. Theorem 3.5, (a) $\Rightarrow$ (c). Since F_0 is a face of P containing x_0, it follows that $y \in F_0$ (and $z \in F_0$). Since $F' \subset H$, this proves $\subset$ in (1). Conversely, let $y \in H \cap F_0$ with $y \neq x_0$. Then there exists a point $z \in F_0$ such that $x_0 \in]y, z[$, cf. Theorem 3.5, (a) $\Rightarrow$ (c). Since x_0 and y are in H, then z must also be in H, whence $y, z \in P'$. Since F' is a face of P' containing x_0, it follows that $y \in F'$ (and $z \in F'$). This proves $\supset$ in (1).

(d) For $F' = \emptyset$, there is nothing to prove. For $F' \neq \emptyset$, we refer to the proof of (c). Let F be any face of P such that $F' = H \cap F$. Then F must contain the point x_0, and hence $F_0 \subset F$ by the definition of F_0. Corollary 5.5 next shows that if $F_0 \subsetneq F$, then $\dim F_0 < \dim F$. We shall complete the proof by showing that

$$\begin{aligned} \dim F_0 &= \dim F \\ &= \dim F' + 1. \end{aligned}$$

Let G be any face of P such that $F' = H \cap G$. Since F' is not a face of P by assumption, statement (b) applied to G gives $\dim G = \dim F' + 1$. Since this applies to both $G = F_0$ and $G = F$, we have the desired conclusion. □

Let x_0 be a vertex of a d-polytope P in $\mathbb{R}^d$ (where $d \ge 1$). Then there is a supporting hyperplane $H(y, \alpha)$ of P such

$$H(y, \alpha) \cap P = \{x_0\}.$$

Assuming that $P \subset K(y, \alpha)$, it then follows that for some $\beta < \alpha$, all the vertices of P except x_0 are in $K(y, \beta) \setminus H(y, \beta)$, whereas x_0 is in $\mathbb{R}^d \setminus K(y, \beta)$. In other words, there is a hyperplane H which separates x_0 from the remaining vertices of P in the sense that x_0 is on one side of H and the remaining vertices are on the other side of H. By a *vertex-figure* of P at x_0 we mean a set $H \cap P$ (in fact, a $(d-1)$-polytope, cf. Theorem 11.2), where H is a hyperplane separating x_0 from the remaining vertices of P.

When x_0 is a vertex of P, we denote by $\mathscr{F}(P/x_0)$ the set of faces of P containing x_0. It is clear that $(\mathscr{F}(P/x_0), \subset)$ is a sublattice of $(\mathscr{F}(P), \subset)$. This lattice is "essentially" the face-lattice of the vertex-figures of P at x_0:

Theorem 11.2. *Let P be a d-polytope in $\mathbb{R}^d$, let x_0 be a vertex of P, and let $P' = H \cap P$ be a vertex-figure of P at x_0. Then P' is a $(d-1)$-polytope. Furthermore, the mapping*

$$F \mapsto F' := H \cap F$$

is an isomorphism from the sublattice $(\mathscr{F}(P/x_0), \subset)$ of $(\mathscr{F}(P), \subset)$ onto the lattice $(\mathscr{F}(P'), \subset)$.

PROOF. The hyperplane H intersects P and is not a supporting hyperplane. Therefore,

$$H \cap \operatorname{int} P \neq \varnothing$$

by Theorem 4.1. Then P' is a $(d-1)$-polytope by Theorem 11.1(a).

It follows from Theorem 11.1(b) that the mapping

$$F \mapsto F' := H \cap F$$

maps $\mathscr{F}(P/x_0)$ into $\mathscr{F}(P')$, and it follows from Theorem 11.1(d) that the mapping is one-to-one and onto. Finally, it is trivial that the mapping preserves inclusions. □

Corollary 11.3. *Let P be a d-polytope in $\mathbb{R}^d$, and let x_0 be a vertex of P. Then any two vertex-figures of P at x_0 are equivalent.*

PROOF. In fact, if P' and P'' are vertex-figures of P at x_0, then $(\mathscr{F}(P'), \subset)$ and $(\mathscr{F}(P''), \subset)$ are both isomorphic to $(\mathscr{F}(P/x_0), \subset)$, and therefore mutually isomorphic. □

Of course, Corollary 11.3 can also be proved by an argument based directly on Theorem 11.1.

Now, let F_1 and F_2 be faces of a d-polytope P such that $F_1 \subset F_2$. Then the set of faces F of P such that

$$F_1 \subset F \subset F_2$$

will be denoted by $\mathscr{F}(F_2/F_1)$. When $F_1 = \{x_0\}$ and $F_2 = P$, we recover $\mathscr{F}(P/x_0)$. It is clear that $(\mathscr{F}(F_2/F_1), \subset)$ is a sublattice of $(\mathscr{F}(P), \subset)$. It follows from Theorem 11.2 above that in the particular case where F_1 is a vertex and $F_2 = P$, the lattice $(\mathscr{F}(F_2/F_1), \subset)$ can be "realized" as the face-lattice of a suitable polytope. This is true in general (except, of course, when $F_1 = F_2$):

Theorem 11.4. *Let P be a polytope, let F_1 be a j-face of P, and let F_2 be a k-face of P such that $F_1 \subsetneqq F_2$. Then there is a $(k-1-j)$-polytope Q such that $(\mathscr{F}(F_2/F_1), \subset)$ is isomorphic to $(\mathscr{F}(Q), \subset)$. Furthermore, for every isomorphism φ from $(\mathscr{F}(F_2/F_1), \subset)$ onto $(\mathscr{F}(Q), \subset)$ we have*

$$\dim \varphi(F) = \dim F - 1 - j$$

for any face $F \in \mathscr{F}(F_2/F_1)$.

PROOF. We consider F_1 as a j-face of the k-polytope F_2; the faces F of P belonging to $\mathscr{F}(F_2/F_1)$ are then the faces F of F_2 such that $F_1 \subset F$. Let G_2 be a dual polytope of F_2, and let ψ be an anti-isomorphism from $(\mathscr{F}(F_2), \subset)$ onto $(\mathscr{F}(G_2), \subset)$. The faces of G_2 corresponding under ψ to the faces F of F_2 with $F_1 \subset F$ are then the faces $G = \psi(F)$ of G_2 such that

$$\emptyset = \psi(F_2) \subset G = \psi(F) \subset \psi(F_1).$$

This shows that the restriction of ψ to $(\mathscr{F}(F_2/F_1), \subset)$ is an anti-isomorphism from $(\mathscr{F}(F_2/F_1), \subset)$ onto the face-lattice $(\mathscr{F}(\psi(F_1)), \subset)$ of the polytope $\psi(F_1)$. Therefore, if we take Q to be any dual polytope of $\psi(F_1)$, we see that $(\mathscr{F}(F_2/F_1), \subset)$ is isomorphic to $(\mathscr{F}(Q), \subset)$.

To determine the dimension of Q, note that

$$\dim Q = \dim \psi(F_1)$$

by Theorem 10.3. But

$$\begin{aligned} \dim \psi(F_1) &= k - 1 - \dim F_1 \\ &= k - 1 - j \end{aligned}$$

by the dimension formula of Theorem 10.3. Hence,

$$\dim Q = k - 1 - j,$$

as desired.

Finally, let φ be any isomorphism from $(\mathscr{F}(F_2/F_1), \subset)$ onto $(\mathscr{F}(Q), \subset)$. Every face $F \in \mathscr{F}(F_2/F_1)$ is a member of a chain

$$F_1 = G_j \subsetneqq \cdots \subsetneqq G_i \subsetneqq \cdots \subsetneqq G_k = F_2$$

of faces $G_i \in \mathscr{F}(F_2/F_1)$ with

$$\dim G_i = i, \qquad i = j, \ldots, k,$$

cf. Corollary 9.7. Application of φ yields the chain

$$\emptyset = \varphi(F_1) = \varphi(G_j) \subsetneqq \cdots \subsetneqq \varphi(G_i) \subsetneqq \cdots \subsetneqq \varphi(G_k) = \varphi(F_2) = Q.$$

This implies

$$\begin{aligned} -1 &= \dim \varphi(G_j) < \cdots < \dim \varphi(G_i) < \cdots < \dim \varphi(G_k) \\ &= \dim Q = k - 1 - j, \end{aligned}$$

where we have used Corollary 5.5 and the expression for $\dim Q$ found above. This in turn enforces

$$\dim \varphi(G_i) = i - 1 - j, \qquad i = j, \ldots, k,$$

whence

$$\dim \varphi(G_i) = \dim G_i - 1 - j, \qquad i = j, \ldots, k.$$

This completes the proof. □

We note the following:

Theorem 11.5. *Let P and Q be dual d-polytopes, and let*

$$\psi: (\mathscr{F}(P), \subset) \to (\mathscr{F}(Q), \subset)$$

be an anti-isomorphism. Let x_0 be a vertex of P, and let P' be a vertex-figure of P at x_0. Then the facet $\psi(\{x_0\})$ of Q is a dual of P'.

PROOF. We know by Theorem 11.2 that $(\mathscr{F}(P'), \subset)$ and $(\mathscr{F}(P/x_0), \subset)$ are isomorphic. Taking $F_1 = \{x_0\}$ and $F_2 = P$ in the proof of Theorem 11.4, we see that $(\mathscr{F}(P/x_0), \subset)$ is isomorphic to $(\mathscr{F}(R), \subset)$, where R is any polytope dual to $\psi(\{x_0\})$. (This polytope R is denoted by Q in the proof of Theorem 11.4.) Therefore, $(\mathscr{F}(P/x_0), \subset)$ is anti-isomorphic to $(\mathscr{F}(\psi(\{x_0\})), \subset)$, as desired. □

Let x_0 be a vertex of a d-polytope P. Vertex-figures of P at x_0 arise from hyperplanes H separating x_0 from all the remaining vertices of P. We shall prove that in order to have H separating x_0 from all the remaining vertices of P it suffices to have H separating x_0 from those vertices of P that are adjacent to x_0. We first prove:

Theorem 11.6. *Let P be a d-polytope in $\mathbb{R}^d$, let x_0 be a vertex of P, and let $x_1, \ldots, x_k$ be the vertices of P adjacent to x_0. Let $H(y, \alpha)$ be a hyperplane in $\mathbb{R}^d$ such that $x_0 \in H(y, \alpha)$ and $x_1, \ldots, x_k \in K(y, \alpha)$. Then $P \subset K(y, \alpha)$ i.e. $H(y, \alpha)$ is a supporting hyperplane of P. If, in addition, we have $x_1, \ldots, x_k \notin H(y, \alpha)$, then $H(y, \alpha) \cap P = \{x_0\}$.*

PROOF. Let $P' = H' \cap P$ be a vertex-figure of P at x_0, determined by a hyperplane H' separating x_0 from the remaining vertices of P. Theorem 11.1(b), (d)—or Theorem 11.2—tells that the vertices of P' are the 1-point sets $[x_0, x_i] \cap H'$, $i = 1, \ldots, k$. Since both x_0 and $x_1, \ldots, x_k$ are in $K(y, \alpha)$ by assumption, it follows that the vertices of P' are in $K(y, \alpha)$, and, therefore, P' is in $K(y, \alpha)$.

Let x be any vertex of P with $x \neq x_0$. Then x and x_0 are on opposite sides of the hyperplane H', whence $[x_0, x] \cap H'$ is a 1-point set, say $\{x'\}$. Since $x' \in P'$ and $P' \subset K(y, \alpha)$, it follows that $x' \in K(y, \alpha)$. This, in turn, clearly implies $x \in K(y, \alpha)$. In other words, all the vertices of P are in $K(y, \alpha)$, whence $P \subset K(y, \alpha)$.

If, in addition, $x_1, \ldots, x_k$ do not lie in $H(y, \alpha)$, then $]x_0, x_i[\subset \operatorname{int} K(y, \alpha)$ for $i = 1, \ldots, k$. So, all the vertices of P' belong to $\operatorname{int} K(y, \alpha)$, and therefore $P' \subset \operatorname{int} K(y, \alpha)$. For any vertex x of P, $x \neq x_0$, we then have $x' \in \operatorname{int} K(y, \alpha)$, implying that $x \in \operatorname{int} K(y, \alpha)$. (Here, as above, x' denotes the single point in $[x_0, x] \cap H'$.) This shows that the only vertex of P in the exposed face $H(y, \alpha) \cap P$ is x_0, implying that $H(y, \alpha) \cap P = \{x_0\}$, cf. Theorem 7.3. □

Corollary 11.7. *Let P be a d-polytope in $\mathbb{R}^d$, let x_0 be a vertex of P, and let $x_1, \ldots, x_k$ be the vertices of P adjacent to x_0. Then*

$$\operatorname{aff}\{x_0, x_1, \ldots, x_k\} = \mathbb{R}^d.$$

PROOF. If the desired conclusion is not valid, then there is a hyperplane H containing $x_0, x_1, \ldots, x_k$. Then both of the two closed halfspaces bounded by H contain $x_0, x_1, \ldots, x_k$. By Theorem 11.6 this implies that P is contained in both of these halfspaces, whence $P \subset H$. This contradiction completes the proof. □

We can now prove:

Theorem 11.8. *Let P be a d-polytope in $\mathbb{R}^d$, let x_0 be a vertex of P, and let $x_1, \ldots, x_k$ be the vertices of P adjacent to x_0. Let H be a hyperplane in $\mathbb{R}^d$ separating x_0 from $x_1, \ldots, x_k$. Then H separates x_0 from any other vertex of P, whence $H \cap P$ is a vertex-figure of P at x_0.*

PROOF. Given the vertex x_0 and its adjacent vertices $x_1, \ldots, x_k$, let x be any other vertex of P. Let L denote the line through x_0 and x. We first prove that $]x_0, x[$ intersects the set

$$Q_1 := \operatorname{conv}\{x_1, \ldots, x_k\}.$$

Let H_1 be a hyperplane in $\mathbb{R}^d$ orthogonal to L, and let $\pi\colon \mathbb{R}^d \to H_1$ denote the orthogonal projection. Letting

$$Q_0 := \operatorname{conv}\{x_0, x_1, \ldots, x_k\},$$

it is clear that

$$\pi(Q_0) = \operatorname{conv}\{\pi(x_0), \pi(x_1), \ldots, \pi(x_k)\}.$$

In particular, $\pi(Q_0)$ is a polytope with

$$\operatorname{ext} \pi(Q_0) \subset \{\pi(x_0), \pi(x_1), \ldots, \pi(x_k)\},$$

cf. Theorem 7.2, (a) ⇒ (b). Suppose that $\pi(x_0)$ is a vertex of $\pi(Q_0)$. Then there is a supporting hyperplane H_2 of $\pi(Q_0)$ in H_1 with $H_2 \cap \pi(Q_0) = \{\pi(x_0)\}$, cf. Theorem 7.5. But then $\pi^{-1}(H_2) = \operatorname{aff}(H_2 \cup L)$ is a supporting hyperplane of Q_0 in $\mathbb{R}^d$ with $x_1, \ldots, x_k \notin \pi^{-1}(H_2)$ and $x \in \pi^{-1}(H_2)$, contradicting the second statement of Theorem 11.6. Hence, $\pi(x_0)$ is not a vertex of $\pi(Q_0)$. This implies that

$$\pi(x_0) \in \operatorname{conv}\{\pi(x_1), \ldots, \pi(x_k)\},$$

cf. Theorem 7.2, (b) ⇒ (a). Since

$$\begin{aligned}\operatorname{conv}\{\pi(x_1), \ldots, \pi(x_k)\} &= \pi(\operatorname{conv}\{x_1, \ldots, x_k\})\\ &= \pi(Q_1),\end{aligned}$$

it follows that $\pi(x_0) \in \pi(Q_1)$, implying that L intersects Q_1. However, since x_0 and x are vertices of P, and Q_1 is a subset of P, every point of L in Q_1 must lie between x_0 and x.

To complete the proof, let H be a hyperplane separating x_0 from $x_1, \dots, x_k$, and let K be that closed halfspace bounded by H which contains $x_1, \dots, x_k$. Then $x_1, \dots, x_k$ belong to int K, whence also $Q_1 \subset \text{int } K$. Let x be any other vertex of P. Using what we have proved above, we see that at least one point of $]x_0, x[$ is in int K. By the convexity of $\mathbb{R}^d \setminus \text{int } K$ we must then also have $x \in \text{int } K$, as desired. □

The next theorem has an interesting application in the proof of Theorem 11.10.

Theorem 11.9. *Let P be a d-polytope in $\mathbb{R}^d$ (where $d \geq 1$), and let x_0 be a vertex of P. Then there is a point $x'_0 \in \text{int } P$ such that the hyperplane through x'_0 with $x'_0 - x_0$ as a normal separates x_0 from the remaining vertices of P.*

Proof. We may assume that $x_0 = o$. Let $x_1, \dots, x_n$ be the remaining vertices of P, and let

$$P' := \text{conv}\{x_1, \dots, x_n\}.$$

Then P' is a polytope with vertices $x_1, \dots, x_n$, and x_0 is not in P'. Let $v \in P'$ be such that

$$\langle v, v \rangle = \min\{\langle x, x \rangle \mid x \in P'\}. \tag{2}$$

The existence of v follows by noting that the mapping

$$x \mapsto \langle x, x \rangle = \|x\|^2$$

is continuous on the compact set P'. (The point v is in fact the unique point of P' nearest to o.) Since $o \notin P'$, it follows that

$$0 < \langle v, v \rangle. \tag{3}$$

We claim that

$$\langle v, v \rangle = \min\{\langle v, x \rangle \mid x \in P'\}. \tag{4}$$

(Hence $H(v, \alpha)$ with $\alpha := \langle v, v \rangle$ is a supporting hyperplane of P' at v.) To see this, let $x \in P'$ and let $\lambda \in]0, 1[$. Then $\lambda x + (1 - \lambda)v$ is in P', whence by (2)

$$\begin{aligned} \langle v, v \rangle &\leq \langle \lambda x + (1 - \lambda)v, \lambda x + (1 - \lambda)v \rangle \\ &= \langle v, v \rangle + 2\lambda(\langle v, x \rangle - \langle v, v \rangle) + \lambda^2 \langle v - x, v - x \rangle. \end{aligned}$$

Re-arranging and dividing by 2λ yields

$$\langle v, v \rangle - \langle v, x \rangle \leq (\lambda/2)\langle v - x, v - x \rangle.$$

This holds for $\lambda \in]0, 1[$. By continuity it must also hold for $\lambda = 0$, i.e. (4) holds. Now, (3) and (4) imply

$$\langle v, x_i \rangle > 0, \qquad i = 1, \dots, n.$$

By continuity we then have

$$\langle u, x_i \rangle > 0, \qquad i = 1, \dots, n \tag{5}$$

for all u belonging to some ball $B(v, \varepsilon)$. In particular, o is not in $B(v, \varepsilon)$. Let u_0 be a point in $B(v, \varepsilon) \cap \operatorname{int} P$, cf. Theorem 3.4(c). Then $H(u_0, 0)$ is a hyperplane through o with all the vertices $x_1, \ldots, x_n$ strictly on one side. Therefore, for λ sufficiently small, $0 < \lambda < 1$, the hyperplane H parallel to $H(u_0, 0)$ through $x_0' := \lambda u_0$ separates o from $x_1, \ldots, x_n$. Finally, it is clear that H has $x_0' - x_0$ $(= x_0' = \lambda u_0)$ as a normal, and it follows from Theorem 3.3 that $x_0' \in \operatorname{int} P$, since $x_0' \in]o, u_0[$. □

The next theorem is an application of Theorem 11.9. The theorem illustrates how the polar operation can be used to produce polytopes equivalent to a given polytope with desirable properties. The proof is based on the observation that if P and $x + P$ both have o as an interior point, then P° and $(x + P)^\circ$ must be equivalent since P° is a dual of P, $(x + P)^\circ$ is a dual of $x + P$, and P and $x + P$ are equivalent. Theorem 11.10 is needed in Section 19.

Theorem 11.10. *Let P be a d-polytope in $\mathbb{R}^d$, and let F be a facet of P. Then there is a d-polytope P_1 in $\mathbb{R}^d$ equivalent to P such that the orthogonal projection of $\mathbb{R}^d$ onto the hyperplane spanned by the facet F_1 of P_1 corresponding to the facet F of P maps $P_1 \backslash F_1$ into* ri F_1.

PROOF. We may assume that $o \in \operatorname{int} P$. Let $Q := P^\circ$; then $Q^\circ = P$. Let y_0 be the vertex of Q conjugate to F, cf. Theorem 9.8. Let $y_1, \ldots, y_n$ be the remaining vertices of Q. Use Theorem 11.9 to get $y_0' \in \operatorname{int} Q$ such that

$$\langle y_i, y_0 - y_0' \rangle < \langle y_0', y_0 - y_0' \rangle < \langle y_0, y_0 - y_0' \rangle, \qquad i = 1, \ldots, n.$$

Take $Q_1 := Q - y_0'$; then Q_1 is a d-polytope with o in its interior. The vertices of Q_1 are the points $y_i - y_0'$, $i = 0, \ldots, n$. Take $P_1 := Q_1^\circ$. Since Q and Q_1 are equivalent, it follows that P and P_1 are equivalent (under an obvious lattice isomorphism). The facet F_1 of P_1 corresponding to F, of course, is the facet of P_1 conjugate to the vertex $y_0 - y_0'$ of Q_1. Hence,

$$\operatorname{aff} F_1 = H(y_0 - y_0', 1).$$

For $x \in \mathbb{R}^d$, the orthogonal projection of x onto aff F_1 is the point $x' = x + \lambda(y_0 - y_0')$, where λ is determined by

$$\begin{aligned} 1 &= \langle x', y_0 - y_0' \rangle \\ &= \langle x, y_0 - y_0' \rangle + \lambda \langle y_0 - y_0', y_0 - y_0' \rangle, \end{aligned}$$

i.e.

$$\lambda = \frac{1 - \langle x, y_0 - y_0' \rangle}{\|y_0 - y_0'\|^2}.$$

Now, by Theorem 9.1(a),

$$P_1 = \bigcap_{i=0}^{n} K(y_i - y_0', 1).$$

Hence, for $x \in P_1$ we have $\langle x, y_i - y'_0 \rangle \leq 1$ for $i = 0, \ldots, n$, and we have $\langle x, y_0 - y'_0 \rangle < 1$ for $x \in P_1 \setminus F_1$. For $x \in P_1 \setminus F_1$ and x' as above we then get for $i = 1, \ldots, n$,

$$\begin{aligned}\langle x', y_i - y'_0 \rangle &= \langle x, y_i - y'_0 \rangle + \lambda \langle y_0 - y'_0, y_i - y'_0 \rangle \\ &\leq 1 + \lambda \langle y_0 - y'_0, y_i - y'_0 \rangle \\ &< 1,\end{aligned}$$

since $\lambda > 0$ and $\langle y_0 - y'_0, y_i - y'_0 \rangle < 0$. This shows that

$$x' \in \operatorname{int} K(y_i - y'_0, 1)$$

for $i = 1, \ldots, n$, whence

$$x' \in \operatorname{int} \bigcap_{i=1}^{n} K(y_i - y'_0, 1).$$

Since

$$F_1 = H(y_0 - y'_0, 1) \cap \bigcap_{i=1}^{n} K(y_i - y'_0, 1),$$

it follows that $x' \in \operatorname{ri} F_1$, as desired. □

In Theorem 11.1 we described in great detail the facial structure of a polytope of the form $P' = H \cap P$, where H is a hyperplane and P is a polytope whose interior is intersected by H. In a similar way we can describe the facial structure of a polytope of the form $P' = K \cap P$, where K is a closed halfspace and P is a polytope whose interior is intersected by the hyperplane H bounding K. We mention a particular case.

Theorem 11.11. *Let P be a d-polytope in $\mathbb{R}^d$, let H be a hyperplane in $\mathbb{R}^d$ with*

$$H \cap \operatorname{int} P \neq \emptyset, \qquad H \cap \operatorname{ext} P = \emptyset,$$

and let K be one of the two closed halfspaces bounded by H. Then we have:

(a) *The set $P' := K \cap P$ is a d-polytope, and $H \cap P$ is a facet of P'.*
(b) *Let F be a face of P such that $K \cap F \neq \emptyset$. Then $F' := K \cap F$ is a face of P', and $\dim F' = \dim F$.*
(c) *Let F' be a face of P'. Then either F' is a face of the facet $H \cap P$, or there is a unique face F of P such that $F' = K \cap F$.*

Proof. (a) This is obvious, cf. Corollary 9.4.

(b) It is obvious that F' is a face of P'. If $F \subset K$, then the dimension formula is trivial. If $F \not\subset K$, then there must be points of F on both sides of H; for if not, then H would be a supporting hyperplane of F with $F \subset (\mathbb{R}^d \setminus K) \cup H$, contradicting the assumption that H contains no vertex of P. But then H must intersect $\operatorname{ri} F$, cf. Theorem 4.1, (a) $\Rightarrow$ (b). This in turn clearly implies $\dim F' = \dim F$.

(c) Suppose that F' is not a face of $H \cap P$. Then F' is not a subset of $H \cap P$, cf. Theorem 5.2, whence $H \cap \text{ri } F' = \emptyset$, cf. Theorem 4.1, (b) $\Rightarrow$ (a).

We first prove uniqueness of F. If F_1 and F_2 were faces of P such that $F' = K \cap F_1 = K \cap F_2$, then we would also have $F' = K \cap (F_1 \cap F_2)$. By (b) we would then have $\dim F_1 = \dim F_2 = \dim(F_1 \cap F_2)$, implying that $F_1 = F_2$, cf. Theorem 5.2 and Corollary 5.5.

To prove existence of F, take $x_0 \in \text{ri } F'$; then $x_0 \in P \cap \text{int } K$ since $H \cap \text{ri } F' = \emptyset$. Let H' be a supporting hyperplane of P' such that $H' \cap P' = F'$. Then H' is also a supporting hyperplane of P. For if some point y of P not in K were on the wrong side of H', then the entire segment $]x_0, y]$ would be on the wrong side; but $]x_0, y]$ contains a whole segment of points from $K \cap P = P'$, whence H' could not be a supporting hyperplane of P'. Hence, H' is a supporting hyperplane of P. But then $F := H' \cap P$ is a face of P with $K \cap F = F'$, as desired. □

If in Theorem 11.11 the set of vertices of P not in K are the vertices of a face F, then the polytope P' is said to be obtained from P by *truncation* of the face F. The operation of truncation produces one "new" facet. The old facets of P all "survive", except of course F itself, if F is a facet. The dual operation of truncating a facet is called *pulling a vertex*. It consists in taking the convex hull of the polytope and one "new" vertex (outside the polytope) such that one "old" vertex disappears. The dual operation of truncating a vertex is that of *adding a pyramid over one of the facets*. A precise description of the duality can be given in terms of polarity as explained in Theorem 9.1.

Exercise

11.1. Let F_1, F_2, and F_3 be faces of a polytope P such that $F_1 \subsetneqq F_2 \subsetneqq F_3$. Let Q be a polytope such that $(\mathscr{F}(F_3/F_1), \subset)$ is isomorphic to $(\mathscr{F}(Q), \subset)$ under the isomorphism φ. Verify that $(\mathscr{F}(F_2/F_1), \subset)$ is isomorphic to $(\mathscr{F}(\varphi(F_2)), \subset)$, and $(\mathscr{F}(F_3/F_2), \subset)$ is isomorphic to $(\mathscr{F}(Q/\varphi(F_2)), \subset)$.

§12. Simple and Simplicial Polytopes

In this section we introduce two important classes of polytopes, namely, the simple polytopes and the simplicial polytopes. Both classes are defined by "non-degeneracy" conditions; actually, the conditions are dual. The "non-degeneracy" makes these polytopes much easier to handle than polytopes in general; in fact, with one important exception, the combinatorial theory to be developed in Chapter 3 deals only with simple and simplicial polytopes.

Because of the duality there is no formal reason to prefer one of the two classes to the other. However, certain problems are treated most conveniently in terms of simple polytopes.

We remind the reader that an e-polytope is an e-simplex provided that its vertices form an affinely independent $(e + 1)$-family, cf. Sections 2 and 7. We begin with a discussion of the facial structure of simplices.

Theorem 12.1. *Let S be an e-simplex in $\mathbb{R}^d$, and let F be a proper face of S. Then F is also a simplex.*

PROOF. The vertices of F are those vertices of S which are in F, cf. Theorem 7.3. Any subfamily of an affinely independent family of points is itself affinely independent. Therefore, since F is the convex hull of its vertices, cf. Theorem 7.2(c), it follows that F is a simplex. □

Theorem 12.2. *Let S be an e-simplex in $\mathbb{R}^d$, let X be a non-empty subset of* ext S, *and let* $F :=$ conv X. *Then F is a face of S, and* ext $F = X$.

PROOF. Let ext $S = \{x_1, \ldots, x_{e+1}\}$, and let us assume that $X = \{x_1, \ldots, x_k\}$. To prove that F is a face of S, we shall show that if y_0 and y_1 are two points of S such that for some $t \in]0, 1[$, the point

$$y_t := (1 - t)y_0 + ty_1$$

is in F, then y_0 and y_1 must be in F. Each x in S has a unique representation

$$x = \sum_{i=1}^{e+1}{}^{c} \lambda_i x_i. \tag{1}$$

Points x from S actually belonging to F are characterized by the property that $\lambda_i = 0$ for $i = k + 1, \ldots, e + 1$. Now, we have

$$y_0 = \sum_{i=1}^{e+1}{}^{c} \lambda_{0i} x_i$$

and

$$y_1 = \sum_{i=1}^{e+1}{}^{c} \lambda_{1i} x_i,$$

whence

$$y_t = \sum_{i=1}^{e+1}{}^{c} ((1 - t)\lambda_{0i} + t\lambda_{1i})x_i.$$

But we also have $y_t \in F$, i.e.

$$y_t = \sum_{i=1}^{k}{}^{c} \lambda_{ti} x_i.$$

By the uniqueness of representations (1) we then get

$$(1 - t)\lambda_{0i} + t\lambda_{1i} = 0, \qquad i = k + 1, \ldots, e + 1.$$

This clearly implies

$$\lambda_{0i} = \lambda_{1i} = 0, \qquad i = k + 1, \ldots, e + 1,$$

whence y_0 and y_1 are in F, as desired.

Finally, Theorem 7.2, (a) $\Rightarrow$ (b) shows that

$$\text{ext } F \subset \{x_1, \ldots, x_k\}.$$

The opposition inclusion is clear, cf. the "only if" part of Theorem 5.2. □

The two theorems above contain the basic information about faces of simplices. We have the following corollaries:

Corollary 12.3. *Let S be an e-simplex in $\mathbb{R}^d$, and let F be a j-face of S, where $-1 \leq j \leq e$. Then for $k = j, \ldots, e$, the number of k-faces of S containing F equals*

$$\binom{e - j}{k - j}.$$

Proof. By Theorems 12.1 and 12.2 there is a one-to-one correspondence between the k-faces of S containing F, and the choices of $(k + 1) - (j + 1)$ vertices from the $(e + 1) - (j + 1)$ vertices of S not in F. This proves the assertion. □

Corollary 12.4. *Let S be an e-simplex in $\mathbb{R}^d$. Then for $-1 \leq k \leq e$, the number of k-faces of S equals*

$$\binom{e + 1}{k + 1}.$$

Proof. Take $j = -1$ in Corollary 12.3. □

Corollary 12.5. *Let S be an e-simplex in $\mathbb{R}^d$, and let F be a k-face of S, where $-1 \leq k \leq e$. Then the number of facets of S containing F equals $e - k$.*

Proof. It follows from Corollary 12.3 that there are

$$\binom{e - k}{(e - 1) - k} = e - k$$

facets of S containing a given k-face F. This proves the assertion. □

In Corollary 12.5, note that F is the intersection of the $e - k$ facets containing F, cf. Theorem 10.4. Conversely, the intersection of $e - k$ facets is a k-face:

Corollary 12.6. *Let S be an e-simplex in $\mathbb{R}^d$, and let $F_1, \ldots, F_{e-k}$ be $e - k$ facets of S, where $-1 \leq k \leq e - 1$. Then $F_1 \cap \cdots \cap F_{e-k}$ is a k-face of S.*

Proof. Let $x_1, \ldots, x_{e+1}$ be the vertices of S. By Theorems 12.1 and 12.2, each F_j is the convex hull of certain e of the $e + 1$ vertices. We may assume that

$$F_j = \operatorname{conv}(\{x_1, \ldots, x_{e+1}\} \setminus \{x_j\}), \qquad j = 1, \ldots, e - k.$$

Then a point x in S is in F_j if and only if in the (unique) representation

$$x = \sum_{i=1}^{e+1}{}^{c} \lambda_i x_i$$

we have $\lambda_j = 0$. Therefore, x is in $F_1 \cap \cdots \cap F_{e-k}$ if and only if $\lambda_1 = \cdots = \lambda_{e-k} = 0$, i.e. if and only if x is in the set

$$\operatorname{conv}\{x_{e-k+1}, \ldots, x_{e+1}\}.$$

But this set is a face of S by Theorem 12.2, and its dimension is k by Theorem 12.1. □

Corollary 12.7. *Let S be an e-simplex in $\mathbb{R}^d$, and let T be a dual e-polytope. Then T is also an e-simplex.*

Proof. It follows from Corollary 12.4 that S has $e + 1$ facets. Dually, T has $e + 1$ vertices, cf. Theorem 10.3. But e-polytopes with $e + 1$ vertices are simplices. □

Corollary 12.8. *Let P be an e-polytope in $\mathbb{R}^d$. Then P is an e-simplex if and only if the number of facets of P is $e + 1$.*

Proof. If P is an e-simplex, then P has $e + 1$ facets by Corollary 12.4. Conversely, if P is an e-polytope with $e + 1$ facets, then any dual Q of P is an e-polytope with $e + 1$ vertices, cf. Theorem 10.3. Hence, Q is an e-simplex, and therefore, by Corollary 12.7, P is also an e-simplex. □

We shall move on to the simplicial and simple polytopes.

A d-polytope P is said to be *simplicial* if for $k = 0, \ldots, d - 1$, each k-face of P has precisely $k + 1$ vertices (i.e. each proper face of P is a simplex).

Any simplex is simplicial, cf. Theorem 12.1, but of course there are many other simplicial polytopes.

In the definition of a simplicial polytope it suffices to require that all facets are simplices:

Theorem 12.9. *A d-polytope P is simplicial if (and only if) each facet of P is a simplex.*

Proof. Let F be a proper face of P. By Corollary 9.7 there is a facet G of P containing F. Then F is a face of G, cf. Theorem 5.2, and since G is a simplex by assumption, F is a simplex by Theorem 12.1. □

Let F be a k-face of a d-polytope P, where $0 \leq k \leq d - 1$. Then by Theorem 10.4 there are at least $d - k$ facets of C containing F (and F is the intersection of these facets). A d-polytope P with the property that for $k = 0, \ldots, d - 1$, the number of facets of P containing any k-face of P equals $d - k$, is called a *simple* polytope.

Any simplex is simple, cf. Corollary 12.5, but of course there are many other simple polytopes.

The two notions introduced above are dual:

Theorem 12.10. *Let P and Q be dual d-polytopes. Then P is simple if and only if Q is simplicial.*

PROOF. Let F and G be proper faces of P and Q, respectively, corresponding under some anti-isomorphism from $(\mathcal{F}(P), \subset)$ onto $(\mathcal{F}(Q), \subset)$. Then

$$\dim G = d - 1 - \dim F,$$

cf. Theorem 10.3. Furthermore, by the duality, saying that F is contained in j facets of P is equivalent to saying that G contains j vertices of Q. Therefore, saying that each k-face of P is contained in precisely $d - k$ facets is equivalent to saying that each $(d - 1 - k)$-face of Q has precisely $d - k$ vertices, $k = 0, \ldots, d - 1$, i.e. each proper face of Q is a simplex. This proves the statement. □

The following, in a sense, is a dual of Theorem 12.9:

Theorem 12.11. *A d-polytope P is simple if (and only if) each vertex of P is contained in precisely d facets.*

PROOF. Let Q be a dual of P. If each vertex of P is contained in precisely d facets, then each facet of Q has precisely d vertices, cf. Theorem 10.3. Therefore, each facet of Q is a simplex, whence Q is simplicial by Theorem 12.9. But then P is simple by Theorem 12.10. □

The following characterization of simple polytopes should be compared to Theorem 10.5:

Theorem 12.12. *A d-polytope P is simple if and only if each vertex of P is incident to precisely d edges of P.*

PROOF. Let Q be a dual of P, and let ψ be an anti-isomorphism from $(\mathcal{F}(P), \subset)$ onto $(\mathcal{F}(Q), \subset)$. Let x be a vertex of P. Then the number of edges of P incident to x equals the number of $(d - 2)$-faces of the $(d - 1)$-face $\psi(\{x\})$ of Q, cf. Theorem 10.3. Therefore, the number of edges incident to a vertex of P is d for each vertex of P, if and only if the number of $(d - 2)$-faces of a $(d - 1)$-face of Q is d for each $(d - 1)$-face of Q. A $(d - 1)$-polytope, however, has d facets if and only if it is a simplex, cf. Corollary 12.8. The statement then follows from Theorems 12.10 and 12.9. □

In Theorem 12.12, note that "incident to precisely d edges" is equivalent to "adjacent to precisely d vertices."

Here is one more characterization of simple polytopes:

Theorem 12.13. *A d-polytope P is simple if and only if each vertex-figure of P is a simplex.*

PROOF. Let Q be a dual of P. Then the facets of Q are duals of the vertex-figures of P, cf. Theorem 11.5. The statement then follows from Theorems 12.10 and 12.9. □

We shall next establish some properties of simple polytopes that will be needed later.

Theorem 12.14. *Let P be a simple d-polytope, and let $F_1, \ldots, F_{d-k}$ be $d - k$ facets of P, where $0 \leq k \leq d - 1$. Let*

$$F := \bigcap_{i=1}^{d-k} F_i,$$

and assume that $F \neq \emptyset$. Then F is a k-face of P, and $F_1, \ldots, F_{d-k}$ are the only facets of P containing F.

PROOF. Let Q be a dual of P, and let ψ be an anti-isomorphism from $(\mathscr{F}(P), \subset)$ onto $(\mathscr{F}(Q), \subset)$. By its definition, F is the largest face contained in the F_i's, whence $\psi(F)$ is the smallest face containing the $\psi(F_i)$'s. It follows from Theorem 10.3 that $\psi(F)$ is a proper face and that the $\psi(F_i)$'s are vertices of Q. Then $\psi(F)$ is a simplex, cf. Theorem 12.10, and therefore the $\psi(F_i)$'s must be all the vertices of $\psi(F)$, cf. Theorem 12.2. Since the number of F_i's is $d - k$, we see that

$$\dim \psi(F) = (d - k) - 1.$$

This implies by duality that the F_i's are all the facets of P containing F, and that

$$\dim F = k,$$

cf. Theorem 10.3. □

Theorem 12.15. *Let P be a simple d-polytope. Then every proper face of P is also simple.*

PROOF. Let F be a proper face of P, and let x be a vertex of F. Letting $k := \dim F$, we shall prove that there are precisely k facets of F containing x, cf. Theorem 12.11. Let Q be a dual of P, and let ψ be an anti-isomorphism from $(\mathscr{F}(P), \subset)$ onto $(\mathscr{F}(Q), \subset)$. Then by Theorem 10.3, the number of facets of F containing x equals the number of $(d - 1 - (k - 1))$-faces of Q contained in the facet $\psi(\{x\})$ of Q and containing the $(d - 1 - k)$-face

$\psi(F)$ of Q. Now, by Theorem 12.10, Q is simplicial, whence $\psi(\{x\})$ is a $(d-1)$-simplex. Therefore, we are seeking the number of $(d-k)$-faces of a $(d-1)$-simplex containing a given $(d-1-k)$-face of that simplex. Corollary 12.3 tells that this number is

$$\binom{(d-1)-(d-1-k)}{(d-k)-(d-1-k)} = \binom{k}{1} = k,$$

as desired. □

Theorem 12.16. *Let P be a simple d-polytope. Then for $0 \le j \le k \le d$ there are precisely*

$$\binom{d-j}{d-k}$$

k-faces of P containing a given j-face of P.

PROOF. For $k = d$, there is nothing to prove. For $k < d$, let Q be a dual of P, and let ψ be an anti-isomorphism from $(\mathscr{F}(P), \subset)$ onto $(\mathscr{F}(Q), \subset)$. Let F be a given j-face of P. Then $\psi(F)$ is a $(d-1-j)$-face of Q, and the number of k-faces of P containing F equals the number of $(d-1-k)$-faces of $\psi(F)$, cf. Theorem 10.3. By Theorem 12.10, $\psi(F)$ is a simplex. The desired number therefore equals

$$\binom{(d-1-j)+1}{(d-1-k)+1} = \binom{d-j}{d-k},$$

cf. Corollary 12.4. □

Theorem 12.17. *Let P be a simple d-polytope, let x_0 be a vertex of P, let $x_1, \ldots, x_k$ be certain k vertices of P adjacent to x_0, and let F be the smallest face of P containing $[x_0, x_1], \ldots, [x_0, x_k]$. Then the following holds:*

(a) $\dim F = k$.

(b) $[x_0, x_1], \ldots, [x_0, x_k]$ *are the only edges of F incident to x_0.*

PROOF. Let Q be a dual of P, and let ψ be an anti-isomorphism from $(\mathscr{F}(P), \subset)$ onto $(\mathscr{F}(Q), \subset)$. Let $G := \psi(F)$. Then, by duality, G is the largest face of Q contained in the $(d-2)$-faces $\psi([x_0, x_1]), \ldots, \psi([x_0, x_k])$ of the $(d-1)$-face $\psi(\{x_0\})$ of Q, cf. Theorem 10.3. Since Q is simplicial, cf. Theorem 12.10, $\psi(\{x_0\})$ is a $(d-1)$-simplex. Corollary 12.6 then shows that

$$\dim G = d - 1 - k.$$

However, by Theorem 10.3 we also have

$$\dim G = d - 1 - \dim F,$$

whence $\dim F = k$, proving (a).

To prove (b), note that by Corollary 12.5 the number of $(d-2)$-faces of the $(d-1)$-face $\psi(\{x_0\})$ containing G is only k. By duality, this means that there are only k edges of F containing x_0. This proves (b). □

Note that once (a) of Theorem 12.17 has been proved, (b) also follows from Theorems 12.15 and 12.12.

In Theorem 12.17, note that F may also be described as the smallest face of P containing $x_0, x_1, \ldots, x_k$, and (b) is equivalent to saying that $x_1, \ldots, x_k$ are the only vertices of F adjacent to x_0.

Theorem 12.18. *Let P be a simple d-polytope in $\mathbb{R}^d$, and let x_0 be a vertex of P. Let P' be a d-polytope obtained from P by truncating the vertex x_0. Then P' is also a simple d-polytope. Moreover,*

$$f_0(P') = f_0(P) + d - 1$$

and

$$f_j(P') = f_j(P) + \binom{d}{j+1}, \qquad j = 1, \ldots, d-1.$$

Proof. Let K denote the closed halfspace in $\mathbb{R}^d$ such that $P' = K \cap P$, and let H denote the bounding hyperplane of K.

To see that P' is simple, we shall show that each vertex of P' is incident to precisely d edges of P', cf. Theorem 12.12. A vertex x of P' is either a vertex of the facet $H \cap P$, or a vertex of P not in H, cf. Theorem 11.11(d). If the latter holds, then there are precisely d edges of P' incident to x by Theorems 12.12 and 11.11(c), (d). If the former holds, then x is the point where a certain edge F of P crosses H, cf. Theorem 11.1(d). The edges of P' incident to x are then the edge $K \cap F$ plus the edges of $H \cap P$ incident to x. But $H \cap P$ is a $(d-1)$-simplex by Theorem 12.13, and therefore the number of edges of $H \cap P$ incident to x equals $d - 1$, cf. Corollary 12.3.

As already noted, $H \cap P$ is a $(d-1)$-simplex. Therefore, the number of j-faces of $H \cap P$ equals

$$\binom{d}{j+1}, \qquad j = 0, \ldots, d-1,$$

cf. Corollary 12.4. The expressions for $f_j(P')$, $j = 0, \ldots, d-1$, then follows by easy applications of Theorem 11.11. □

A d-simplex is both simple and simplicial. We conclude this section by proving that the converse is also true when $d \neq 2$. (The statement is trivially true for $d = 0, 1$. Any 2-polytope is simple and simplicial, but not all 2-polytopes are simplices. Therefore, the statement is not true for $d = 2$.)

Theorem 12.19. *Let $d \neq 2$, and let P be a d-polytope which is both simple and simplicial. Then P is a simplex.*

PROOF. As noted above, we need only consider $d \geq 3$. Let x_0 be a vertex of P, and let $x_1, \ldots, x_d$ be the vertices of P adjacent to x_0, cf. Theorem 12.12. Let

$$S := \operatorname{conv}\{x_0, x_1, \ldots, x_d\}.$$

Corollary 11.7 implies that S is a d-simplex with vertices $x_0, x_1, \ldots, x_d$. Let x_i and x_j be any two of the vertices $x_1, \ldots, x_d$, and let F be the smallest face of P containing $[x_0, x_i]$ and $[x_0, x_j]$. Then F is a 2-face by Theorem 12.17(a). Moreover, since P is simplicial, F is a simplex. In other words, F is a triangle, and since x_0, x_i and x_j are vertices of F, cf. Theorem 5.2, it follows that $\operatorname{ext} F = \{x_0, x_i, x_j\}$. In particular, $[x_i, x_j]$ is an edge of F, and therefore $[x_i, x_j]$ is also an edge of P, cf. Theorem 5.2. Now, let K be a supporting halfspace of S, and let x_k be a vertex of S in the bounding hyperplane H of K. Then trivially all the d edges of S incident to x_k are in K. But as these edges are also edges of P (as we have proved above), and the number of edges of P incident to x_k equals d by Theorem 12.12, it follows that all edges of P incident to x_k are in K. Application of Theorem 11.6 next shows that K is also a supporting halfspace of P. Hence, every supporting halfspace of S also supports P. Since S is the intersection of its supporting halfspaces, cf. Theorem 4.5, it follows that $P \subset S$. On the other hand, it is clear that $S \subset P$, whence $P = S$, showing that P is a simplex. □

EXERCISES

12.1. Give a direct proof of Theorem 12.2 when X is an e-subset of the $(e + 1)$-set ext S. Apply this result to prove Theorem 12.2 by induction.

12.2. Show by counting incidences of vertices and edges that we have $df_0(P) = 2f_1(P)$ for any simple d-polytope P. (This relation is equivalent to the Dehn–Sommerville Relation corresponding to $i = 1$, cf. Theorem 17.1.)

12.3. Let F be a face of a simple d-polytope P. Show that the facets of F are the faces $F \cap G$ such that G is a facet of P with $F \cap G \neq \emptyset$ and $F \not\subset G$.

12.4. Let P be an arbitrary d-polytope, and let P' be a d-polytope obtained by successive truncations of all the facets of P. Show that P' is simple. Verify that $f_{d-1}(P') = f_{d-1}(P)$ and $f_j(P') \geq f_j(P)$ for $j = 0, \ldots, d - 2$. Show that if some k-face F of P is contained in more than $d - k$ facets, then $f_j(P') > f_j(P)$ for $j = 0, \ldots, k + 1$.

12.5. A d-polytope P is said to be *k-simplicial* if each k-face of P is a simplex, and *k-simple* if each $(d - 1 - k)$-face is contained in precisely $k + 1$ facets.

Verify the following: If P and Q are dual, then P is k-simplicial if and only if Q is k-simple. Every d-polytope is 0-simplicial, 1-simplicial, 0-simple and 1-simple. A d-polytope is simplicial or simple if and only if it is $(d - 1)$-simplicial or $(d - 1)$-simple, respectively. If P is k-simplicial or k-simple, then P is also h-simplicial or h-simple, respectively, for $h < k$.

Prove that if a d-polytope P is k_1-simplicial and k_2-simple with $k_1 + k_2 \geq d + 1$, then P is a simplex.

§13. Cyclic Polytopes

It is easy to see that the 3-simplices are the only 3-polytopes with the property that any two vertices are adjacent. Surprisingly enough, the same statement with 3 replaced by any $d \geq 4$ is not true; counter-examples are provided by polytopes of the type to be introduced in this section.

For $d \geq 2$, the *moment curve* $\mathcal{M}_d$ in $\mathbb{R}^d$ is the curve parametrized by

$$t \mapsto x(t) := (t, t^2, \ldots, t^d), \qquad t \in \mathbb{R}.$$

This curve has the following interesting property:

Theorem 13.1. *Any hyperplane H in $\mathbb{R}^d$ contains at most d points from $\mathcal{M}_d$.*

PROOF. Let $H = H(y, \alpha)$, where

$$y = (\beta_1, \ldots, \beta_d).$$

Then $x(t) \in H(y, \alpha)$ if and only if

$$\beta_1 t + \cdots + \beta_d t^d = \alpha.$$

By the Fundamental Theorem of Algebra there are at most d values of t satisfying this equation, which proves the assertion. □

Corollary 13.2. *Let $t_1, \ldots, t_n$ be distinct real numbers, where $n \leq d + 1$. Then the n-family $(x(t_1), \ldots, x(t_n))$ of points from $\mathbb{R}^d$ is affinely independent.*

PROOF. If $(x(t_1), \ldots, x(t_n))$ is affinely dependent, then all the points $x(t_1), \ldots, x(t_n)$ belong to some affine subspace A with $\dim A \leq n - 2$. If $n < d + 1$, choose $t_{n+1}, \ldots, t_{d+1}$ such that $t_i \neq t_j$ for $i, j = 1, \ldots, d + 1$ and $i \neq j$. Then $x(t_1), \ldots, x(t_{d+1})$ all belong to some affine subspace of dimension at most

$$(n - 2) + (d + 1 - n) = d - 1.$$

This shows in particular that $x(t_1), \ldots, x(t_{d+1})$ all belong to some hyperplane, contradicting Theorem 13.1. □

By a *cyclic polytope* of type $C(n, d)$, where $n \geq d + 1$ and $d \geq 2$, we mean a polytope of the form

$$P = \operatorname{conv}\{x(t_1), \ldots, x(t_n)\},$$

where $t_1, \ldots, t_n$ are distinct real numbers.

Note that a cyclic polytope of type $C(d + 1, d)$ is a d-simplex by Corollary 13.2.

Theorem 13.3. *Let $P = \operatorname{conv}\{x(t_1), \ldots, x(t_n)\}$ be a cyclic polytope of type $C(n, d)$. Then P is a d-polytope.*

PROOF. By Corollary 13.2, any $(d + 1)$-family formed by distinct points $x(t_i)$ is affinely independent. Therefore

$$\operatorname{aff}\{x(t_1), \ldots, x(t_{d+1})\} = \mathbb{R}^d,$$

implying that $\dim P = d$. □

Theorem 13.4. *Let* $P = \operatorname{conv}\{x(t_1), \ldots, x(t_n)\}$ *be a cyclic polytope of type* $C(n, d)$. *Then*

$$\operatorname{ext} P = \{x(t_1), \ldots, x(t_n)\}.$$

PROOF. The inclusion $\subset$ follows from Theorem 7.2, (a) $\Rightarrow$ (b). Conversely, to show that $x(t_i)$ is a vertex of P, consider the polynomium $p(t)$ of degree 2 defined by

$$\begin{aligned} p(t) &:= -(t - t_i)^2 \\ &= -t_i^2 + 2t_i t - t^2. \end{aligned}$$

Let

$$y := (2t_i, -1, 0, \ldots, 0) \in \mathbb{R}^d.$$

Then

$$p(t) = \langle x(t), y \rangle - t_i^2.$$

Since $p(t) \leq 0$ for all $t \in \mathbb{R}$, and $p(t) = 0$ if and only if $t = t_i$, it follows that $K(y, t_i^2)$ is a supporting halfspace of P with

$$H(y, t_i^2) \cap P = \{x(t_i)\},$$

showing that $x(t_i) \in \operatorname{ext} P$. □

Theorem 13.5. *Let* $P = \operatorname{conv}\{x(t_1), \ldots, x(t_n)\}$ *be a cyclic polytope of type* $C(n, d)$. *Then P is simplicial.*

PROOF. Since P is a d-polytope, cf. Theorem 13.3, it suffices to show that any facet of P is a $(d - 1)$-simplex, cf. Theorem 12.9. Let F be a facet of P. Then the vertices of F are certain of the vertices of P, say $x(t_{i_1}), \ldots, x(t_{i_k})$, cf. Theorem 13.4. Then $k \geq d$, with $k = d$ if and only if F is a $(d - 1)$-simplex. Now, note that aff F is a hyperplane containing the k points $x(t_{i_1}), \ldots, x(t_{i_k})$. Theorem 13.1 then shows that $k \leq d$, whence $k = d$, as desired. □

Theorem 13.5 shows that if certain k vertices of a cyclic polytope P form the set of vertices of a face of P, then that face must be a $(k - 1)$-face. In the following we shall describe which sets of k vertices of P are the vertex sets of faces of P.

We need some notation. Let

$$P = \operatorname{conv}\{x(t_1), \ldots, x(t_n)\}$$

by a cyclic polytope of type $C(n, d)$, and assume that

$$t_1 < \cdots < t_n.$$

(This, of course, is no restriction at all.) Let X be a non-empty subset of $\{x(t_1), \ldots, x(t_n)\}$. By a *component* of X we shall mean a non-empty subset Y of X of the form

$$Y = \{x(t_j), x(t_{j+1}), \ldots, x(t_{k-1}), x(t_k)\}$$

such that $x(t_{j-1}) \notin X$ (if $j > 1$) and $x(t_{k+1}) \notin X$ (if $k < n$). A component Y is called a *proper component* if $x(t_1) \notin Y$ and $x(t_n) \notin Y$. A component containing an even (or odd) number of points is called an *even* (or *odd*) *component.*

With this notation we can now handle the case $k = d$; the remaining values of k will be treated below. The condition of the theorem is known as *Gale's Evenness Condition.*

Theorem 13.6. *Let $P = \operatorname{conv}\{x(t_1), \ldots, x(t_n)\}$ be a cyclic polytope of type $C(n, d)$, where $t_1 < \cdots < t_n$. Let X be a subset of $\{x(t_1), \ldots, x(t_n)\}$ containing d points. Then X is the set of vertices of a facet of P if and only if all proper components of X are even.*

PROOF. Let

$$X = \{x(t_{i_1}), \ldots, x(t_{i_d})\},$$

and note that aff X is a hyperplane by Corollary 13.2. Then Theorem 13.1 shows that $x(t_{i_1}), \ldots, x(t_{i_d})$ are the only vertices of P in aff X. Let

$$p(t) := -\prod_{\nu=1}^{d} (t - t_{i_\nu}).$$

Then $p(t)$ is a polynomium of degree d, and therefore there are real numbers $a_0, a_1, \ldots, a_d$ (with $a_d = -1$) such that

$$p(t) = a_0 + a_1 t + \cdots + a_d t^d.$$

Let

$$y := (a_1, \ldots, a_d).$$

Then

$$p(t) = \langle x(t), y \rangle + a_0.$$

Since

$$p(t_{i_1}) = \cdots = p(t_{i_d}) = 0,$$

we see that

$$x(t_{i_1}), \ldots, x(t_{i_d}) \in H(y, -a_0)$$

whence

$$H(y, -a_0) = \operatorname{aff} X.$$

Now, saying that X is the set of vertices of some facet of P is equivalent to saying that there is a supporting hyperplane H of P such that

$$H \cap \text{ext } P = X.$$

But we have seen that $H(y, -a_0)$ is the only hyperplane containing X. Therefore, X is the set of vertices of a facet of P if and only if $H(y, -a_0)$ supports P, i.e. if and only if all points from the set

$$(\text{ext } P)\backslash X = \{x(t_1), \ldots, x(t_n)\}\backslash\{x(t_{i_1}), \ldots, x(t_{i_d})\}$$

are on the same side of $H(y, -a_0)$.

Suppose that not all the points from $(\text{ext } P)\backslash X$ are on the same side of $H(y, -a_0)$. Then there are, in fact, $x(t_j)$ and $x(t_k)$ from $(\text{ext } P)\backslash X$ such that $t_j < t_k$, all points $x(t_l)$ with $j < l < k$ are in $H(y, -a_0)$, and $x(t_j)$ and $x(t_k)$ are on opposite sides of $H(y, -a_0)$. Saying that $x(t_j)$ and $x(t_k)$ are on opposite sides of $H(y, -a_0)$ is equivalent to saying that $p(t_j)$ and $p(t_k)$ have opposite signs. Now, $p(t)$ changes sign exactly at the values $t = t_{i_1}, \ldots, t_{i_d}$. Therefore, there must be an odd number of values t_{i_v} between t_j and t_k. In other words, the set

$$Y = \{x(t_{j+1}), \ldots, x(t_{k-1})\}$$

is an odd proper component of X. This proves the "if" statement.

To prove the "only if" statement, we reverse the argument above. In fact, suppose that there is an odd proper component of X, say

$$Y = \{x(t_{j+1}), \ldots, x(t_{k-1})\}.$$

Then, by the definition of a proper component, $x(t_j)$ and $x(t_k)$ are in $(\text{ext } P)\backslash X$. Therefore, $p(t_j)$ and $p(t_k)$ are $\neq 0$, and since $p(t)$ changes sign at $t = t_{j+1}, \ldots, t_{k-1}$ when t increases from t_j to t_k, we see that $p(t_j)$ and $p(t_k)$ must have opposite signs, showing that $x(t_j)$ and $x(t_k)$ are on opposite sides of $H(y, -a_0)$. This completes the proof. □

We next use Theorem 13.6 to treat the remaining values of k.

Theorem 13.7. *Let $P = \text{conv}\{x(t_1), \ldots, x(t_n)\}$ be a cyclic polytope of type $C(n, d)$, where $t_1 < \cdots < t_n$. Let X be a subset of $\{x(t_1), \ldots, x(t_n)\}$ containing k points, where $k \leq d$. Then X is the set of vertices of a $(k-1)$-face of P if and only if the number of odd proper components of X is at most $d - k$.*

Proof. The set X is the set of vertices of a face of P if and only if there is a facet G of P such that

$$X \subset \text{ext } G. \tag{1}$$

In fact, if $X = \text{ext } F$ for some face F of P, then by Corollary 9.7 there is a facet G of P containing F, whence (1) holds. Conversely, if (1) holds for a certain facet G, then $X = \text{ext } F$ for some face F of G since G is a simplex, cf. Theorems 13.5 and 12.2; but then F is also a face of P.

Now, by Theorem 13.6, the existence of a facet G such that (1) holds is equivalent to the existence of a $(d-k)$-subset Z of $(\text{ext } P)\setminus X$ such that all proper components of $X \cup Z$ are even. This, in turn, is clearly equivalent to saying that the number of odd proper components of X is at most $d-k$, as desired. □

For small values of k, Theorem 13.7 takes the following form:

Corollary 13.8. *Let $P = \text{conv}\{x(t_1), \ldots, x(t_n)\}$ be a cyclic polytope of type $C(n, d)$, and let k be an integer such that*

$$1 \leq k \leq \lfloor d/2 \rfloor.$$

Then any k of the points $x(t_1), \ldots, x(t_n)$ are the vertices of a $(k-1)$-face of P. Hence,

$$f_{k-1}(P) = \binom{n}{k}.$$

Proof. As in Theorem 13.7, we assume that $t_1 < \cdots < t_n$. When $k \leq \lfloor d/2 \rfloor$, then $d - k \geq k$. Since the number of (odd proper) components of X cannot exceed the number of points in X, the conclusion follows immediately from Theorem 13.7. □

Corollary 13.8 is really striking. It shows, for example, that for any $d \geq 4$ there are d-polytopes P with as many vertices as desired such that any two vertices of P are adjacent.

We conclude this section with the following:

Corollary 13.9. *Let $P = \text{conv}\{x(t_1), \ldots, x(t_n)\}$ and $Q = \text{conv}\{x(s_1), \ldots, x(s_n)\}$ be cyclic polytopes, both of type $C(n, d)$. Then P and Q are equivalent.*

Proof. We may assume that $t_1 < \cdots < t_n$ and $s_1 < \cdots < s_n$. For any face F of P with vertices $x(t_{i_1}), \ldots, x(t_{i_k})$, define

$$\varphi(F) := \text{conv}\{x(s_{i_1}), \ldots, x(s_{i_k})\}.$$

Theorem 13.7 then shows that φ is in fact an isomorphism from $(\mathscr{F}(P), \subset)$ onto $(\mathscr{F}(Q), \subset)$. □

Exercises

13.1. Use Theorem 13.6 to show that for any cyclic polytope P of type $C(n, d)$, the number of facets of P is given by

$$\mu(n, d) := \begin{cases} \dfrac{n}{n - d/2}\dbinom{n - d/2}{n - d}, & d \text{ even;} \\[2ex] 2\dbinom{n - (d+1)/2}{n - d}, & d \text{ odd.} \end{cases}$$

Verify that in both cases,

$$\mu(n, d) = \binom{n - \lfloor (d+1)/2 \rfloor}{n - d} + \binom{n - \lfloor (d+2)/2 \rfloor}{n - d}.$$

(After reading Section 18, this should be compared to the case $j = 0$ in Theorem 18.2.)

13.2. Show that if P is a cyclic polytope of type $C(n, d)$, where d is even, then each vertex-figure of P is equivalent to a cyclic polytope of type $C(n - 1, d - 1)$.

13.3. Verify that if P is a cyclic polytope of type $C(n, d)$ such that each vertex-figure of P is equivalent to a cyclic polytope of type $C(n - 1, d - 1)$, then

$$n \cdot \mu(n - 1, d - 1) = d \cdot \mu(n, d),$$

cf. Exercise 13.1. Use this to show that if d is odd and $n \geq d + 2$, then not every vertex-figure of P is equivalent to a cyclic polytope.

13.4. Give a direct proof of Corollary 13.8 by expanding the idea of the proof of Theorem 13.4.

§14. Neighbourly Polytopes

In Section 13 we met examples of d-polytopes P with the property that for certain values of k, every k-subset of ext P is the set of vertices of a face of P. In this section we shall study general properties of such polytopes.

Let k be a positive integer. We shall say that a d-polytope P with at least $k + 1$ vertices is *k-neighbourly* if every k-subset of ext P is the vertex set of a proper face of P, i.e. conv X is a proper face of P for every k-subset X of ext P. For $d \geq 1$, every d-polytope is 1-neighbourly and every d-simplex is k-neighbourly for all $k \leq d$.

We would like to comment on the condition that P should have at least $k + 1$ vertices. If P has k vertices, then there is only one k-subset of ext P, namely, ext P itself; this set, however, is not the vertex set of a *proper* face. If P has fewer than k vertices, then there are no k-subsets of ext P, and therefore formally every k-subset of ext P *is* the vertex set of a proper face of P. So, without the condition that P should have at least $k + 1$ vertices, P would be k-neighbourly for all $k >$ card(ext P).

For k-neighbourliness of a d-polytope P, only $k \leq d$ is possible. In fact, if $k \geq d + 1$, then (assuming that P has at least k vertices) we can find a k-subset X of ext P such that a certain $(d + 1)$-subset of X forms an affinely independent $(d + 1)$-family; the convex set spanned by X must then have dimension d, and therefore it cannot be the vertex set of a proper face. Actually, we shall prove below (cf. Corollary 14.4) that except for simplices only $k \leq \lfloor d/2 \rfloor$ is possible.

Theorem 13.5 and Corollary 13.8 imply:

Theorem 14.1. *Any cyclic polytope P of type $C(n, d)$ is a simplicial k-neighbourly polytope for all $k \leq \lfloor d/2 \rfloor$.*

We shall next study k-neighbourly polytopes in general.

Theorem 14.2. *Let P be a k-neighbourly d-polytope, and let $1 \leq j < k$. Then P is also j-neighbourly.*

PROOF. Let X be any j-subset of ext P. Since

$$\begin{aligned}\operatorname{card}((\operatorname{ext} P)\backslash X) &\geq d + 1 - j \\ &\geq (k - j) + 1,\end{aligned}$$

we see that for any vertex x of P not in X there is a k-subset $Y(x)$ of ext P with $X \subset Y(x)$ and $x \notin Y(X)$. Let

$$F := \bigcap_{x \in (\operatorname{ext} P)\backslash X} \operatorname{conv} Y(x).$$

Since each conv $Y(x)$ is a face of P, it follows that F is a face of P containing X, but not containing any vertex of P not in X. This shows that

$$\operatorname{ext} F = X,$$

whence P is j-neighbourly. □

Theorem 14.3. *Let P be a k-neighbourly d-polytope, and let X be a subset of* ext P *containing at least $k + 1$ points. Then $Q :=$* conv X *is also k-neighbourly.*

PROOF. Let Y be a k-subset of ext $Q = X$. It follows from the k-neighbourliness of P that the set conv Y is a (proper) face of P. Being a proper subset of Q, it must then also be a proper face of Q. □

The next theorem has important implications.

Theorem 14.4. *Let P be a k-neighbourly d-polytope. Then every face F of P with*

$$0 \leq \dim F \leq 2k - 1$$

is a simplex.

PROOF. Let $j := \dim F$. Suppose that F is not a simplex. Then F has at least $j + 2$ vertices. Let M be a $(j + 2)$-subset of ext F. By Radon's Theorem, Corollary 2.7, there are non-empty complementary subsets M_1 and M_2 of M such that

$$\operatorname{conv} M_1 \cap \operatorname{conv} M_2 \neq \emptyset. \tag{1}$$

At least one of the two sets M_1 and M_2 contains at most k points. In fact, if both contained more than k points, then we would have

$$\begin{aligned} j + 2 &= \operatorname{card} M_1 + \operatorname{card} M_2 \geq (k + 1) + (k + 1) \\ &= 2k + 2 \geq \dim F + 3 \\ &= j + 3, \end{aligned}$$

a contradiction. We may assume that card $M_1 \leq k$. Then by Theorem 14.2 and the k-neighbourliness of P, the set conv M_1 is a proper face of P. Let H be a supporting hyperplane of P such that

$$H \cap P = \operatorname{conv} M_1,$$

and let

$$y \in \operatorname{conv} M_1 \cap \operatorname{conv} M_2,$$

cf. (1). Then, in particular, y is in H, and therefore at least one vertex of conv M_2 must be in H. Such a vertex must then be a vertex of conv M_1, which is contradicted by the fact that M_1 and M_2 are disjoint. This completes the proof. □

Corollary 14.5. *Let P be a k-neighbourly d-polytope, where $\lfloor d/2 \rfloor < k$. Then P is a simplex.*

PROOF. Since $\lfloor d/2 \rfloor < k$ implies $d \leq 2k - 1$, we can apply Theorem 14.4 with $F = P$. □

Note that, as a consequence, the only 2-neighbourly and 3-neighbourly 3-polytopes are the 3-simplices. So, the notion of k-neighbourly d-polytopes is only of real interest for $d \geq 4$.

Corollary 14.6. *Let P be a $(d/2)$-neighbourly d-polytope, where d is even. Then P is simplicial.*

PROOF. Let F be a facet of P. Then dim $F = d - 1 = 2k - 1$ with $k = d/2$. Theorem 14.4 next shows that F is a simplex, whence P is simplicial, cf. Theorem 12.9. □

Theorem 14.7. *A simple d-polytope P is a dual of a k-neighbourly polytope if and only if any k facets of P have a non-empty intersection.*

PROOF. Let Q be a dual of P. Then Q is a d-polytope by Theorem 10.3, and Q is simplicial by Theorem 12.10. By the duality, any k vertices of Q belong to a proper face of Q if and only if any k facets of P have a non-empty intersection. But since Q is simplicial, then any k vertices of Q belonging to a proper face of Q are actually the vertices of a proper face. This proves the statement. □

Theorem 14.1 and Corollary 14.5 show that except for simplices, no polytopes are "more neighbourly" than the cyclic polytopes. In the following, $\lfloor d/2 \rfloor$-neighbourly d-polytopes will simply be called *neighbourly polytopes*. The duals of such polytopes are called the *dual neighbourly polytopes*.

There are neighbourly polytopes other than those equivalent to cyclic polytopes. This is trivial for 3-polytopes since every 3-polytope is neighbourly. However, higher-dimensional examples are known.

Finally, let us remark that for odd $d \geq 3$ there are non-simplicial neighbourly polytopes, cf. Corollary 14.6. In fact, let P be a d-pyramid in $\mathbb{R}^d$ whose basis Q is a neighbourly $(d-1)$-polytope. Then P is neighbourly, cf. Theorem 7.7. On the other hand, if Q is not a simplex, then P is not simplicial.

Exercises

14.1. Let P be a k-neighbourly d-polytope. Show that each vertex-figure of P is $(k-1)$-neighbourly.

14.2. Show that every neighbourly d-polytope is $(d-2)$-simplicial, cf. Exercise 12.5.

§15. The Graph of a Polytope

The vertices and edges of a polytope P form in an obvious way a non-oriented graph which we shall denote by $\mathcal{G}(P)$. (For graph-theoretic notions, see Appendix 2.) In this section we shall obtain information about connectedness properties of $\mathcal{G}(P)$. The proofs will be based on a technique for turning $\mathcal{G}(P)$ into an oriented graph $\mathcal{G}(P, w)$ by means of an "admissible" vector w. This "oriented graph technique" will also be used in later sections.

In the following, let P be a d-polytope in $\mathbb{R}^d$, where $d \geq 1$. A vector $w \in \mathbb{R}^d$ is said to be *admissible* for P if $\langle x, w\rangle \neq \langle y, w\rangle$ for any two vertices x and y of P. Geometrically, this means that no hyperplane in $\mathbb{R}^d$ with w as a normal contains more than one vertex of P.

Concerning the existence of admissible vectors we have:

Theorem 15.1. *For any d-polytope P in $\mathbb{R}^d$, the set of admissible vectors is dense in $\mathbb{R}^d$, i.e. for any $y \in \mathbb{R}^d$ and any $\varepsilon > 0$ there is an admissible vector w with $\|y - w\| < \varepsilon$.*

Proof. We first remark that the union of a finite number of hyperplanes in $\mathbb{R}^d$ has no interior points. This follows by repeated application of the observation that for any non-empty open set O in $\mathbb{R}^d$ and any hyperplane H in $\mathbb{R}^d$, the set $O\setminus H$ is again non-empty and open.

Now, let

$$\text{ext}\, P = \{x_1, \ldots, x_k\}$$

and let

$$V := \{x_j - x_i \mid i, j = 1, \ldots, k;\, i \neq j\}.$$

From the remark above it follows that the open ball

$$\{z \in \mathbb{R}^d \mid \|y - z\| < \varepsilon\}$$

is not contained in the union of the hyperplanes $H(v, 0)$, $v \in V$. In other words, there is a $w \in \mathbb{R}^d$ such that $\|y - w\| < \varepsilon$ and $\langle x_j - x_i, w\rangle \neq 0$ for $i \neq j$. This proves the statement. □

Any vector w which is admissible for P induces an *orientation* of the edges of P according to the following rule: An edge $[x, y]$ is oriented towards x and away from y if

$$\langle x, w\rangle > \langle y, w\rangle.$$

The oriented graph thus defined will be denoted by $\mathscr{G}(P, w)$.

Let w be admissible for P. Calling w the "down direction," we see that the edges of $\mathscr{G}(P, w)$ are oriented "downwards." In the following, we shall maintain this terminology which enables us to state that one vertex is "above" or "below" some other vertex, etc. We can also speak about the "top" vertex and the "bottom" vertex of P.

Given an admissible vector w for P, it is clear that the top vertex of P has in-valence 0 in $\mathscr{G}(P, w)$ and that the bottom vertex of P has out-valence 0 in $\mathscr{G}(P, w)$. We actually have:

Theorem 15.2. *In a graph $\mathscr{G}(P, w)$, the top vertex is the only vertex of P whose in-valance is 0, and the bottom vertex is the only vertex of P whose out-valence is 0.*

PROOF. Let x be a vertex of P whose in-valence is 0. Then all the vertices of P adjacent to x are below x. This implies that there is a hyperplane H with w as a normal such that x is above H and all the vertices adjacent to x are below H. Using Theorem 11.8 we then see that all vertices of P except x are below H, showing that x must be the top vertex.

The statement about the bottom vertex can be proved in a similar manner, or by observing that when w is admissible, then $-w$ is also admissible, and, moreover, the in-valence of a vertex x in $\mathscr{G}(P, -w)$ equals the out-valence of x in $\mathscr{G}(P, w)$. □

Theorem 15.3. *Let P be a d-polytope in $\mathbb{R}^d$, and let F be a proper face of P. Then there is an admissible vector w such that each vertex of F is above each vertex of P not in F.*

PROOF. Let $H(y, \alpha)$ be a supporting hyperplane of P with $H(y, \alpha) \cap P = F$, cf. Theorem 7.5. We may assume that $\langle x, y\rangle \geq \alpha$ for all $x \in P$. Let

$$\gamma := \min\{\langle x', y\rangle \mid x' \in \text{ext } P \backslash \text{ext } F\}$$

and let

$$\delta := \max\{\|x' - x''\| \mid x' \in \text{ext } P \backslash \text{ext } F, x'' \in \text{ext } F\}.$$

Note that $\gamma > \alpha$. Take

$$\varepsilon := (\gamma - \alpha)/2\delta.$$

Now, by Theorem 15.1 there is an admissible vector w such that $\|y - w\| < \varepsilon$. Let $z := y - w$. Then for any $x' \in \text{ext } P \backslash \text{ext } F$ and any $x'' \in \text{ext } F$ we have

$$\begin{aligned}\langle x' - x'', w\rangle &= \langle x' - x'', y - z\rangle \\ &= \langle x', y\rangle - \langle x'', y\rangle - \langle x' - x'', z\rangle \\ &\geq \gamma - \alpha - \|x' - x''\| \, \|z\| \\ &> \gamma - \alpha - \delta\varepsilon = (\gamma - \alpha)/2 \\ &> 0,\end{aligned}$$

from where the statement follows immediately. □

Theorem 15.4. *Let P be a d-polytope in $\mathbb{R}^d$, let F be a proper face of P, and let M be a subset of* ext F. *Then there is an admissible vector w for P such that for any vertex x of P not in M there is a path in $\mathscr{G}(P)$ joining x and the bottom vertex v of $\mathscr{G}(P, w)$ without entering M.*

PROOF. By Theorem 15.3 there is an admissible vector w such that each vertex of F is above each vertex not in F. Let x be any vertex not in M. When x is the bottom vertex v, there is nothing to prove. When $x \neq v$, it follows from Theorem 15.2 that there is at least one edge going downwards from x. Let this edge be $[x, x_1]$. If $x_1 = v$, we have a path from x to v. If $x_1 \neq v$, then Theorem 15.2 takes us one step further down. Continuing this way, we obtain a "descending" path joinint x and v. It remains to be shown that we can stay outside M. Note that once we are outside F, we are below F, and therefore we stay outside M from that point on. So, if we can choose the first edge of the path in such a manner that the edge is not an edge of F, we have the desired conclusion. If x is not in F, this is automatically fulfilled. If x is in F, we apply Corollary 11.7 to see that at least one edge of P with x as an end-point is not in F. This completes the proof. □

Theorem 15.5. *Let P be a d-polytope, let F be a proper face of P, and let M be a (possibly empty) subset of* ext F. *Then the subgraph of $\mathscr{G}(P)$ spanned by* (ext P)$\backslash M$ *is connected. In particular, $\mathscr{G}(P)$ is connected.*

PROOF. Denoting by Γ the subgraph of $\mathscr{G}(P)$ spanned by (ext P)$\backslash M$, Theorem 15.4 shows that there is a vertex v of Γ such that any vertex of Γ can be joined to v by a path in Γ. This implies that any two vertices of Γ can be joined by a path in Γ via v, showing that Γ is connected. Taking $M = \varnothing$ shows that $\mathscr{G}(P)$ is connected. □

Theorem 15.5 showed that $\mathscr{G}(P)$ is connected. A much stronger result is the following:

Theorem 15.6. *Let P be a d-polytope. Then $\mathscr{G}(P)$ is d-connected.*

Proof. It suffices to show that for any set N of $d - 1$ vertices of P, the subgraph Γ spanned by (ext $P)\backslash N$ is connected, cf. Appendix 2, Theorem A2.1. If there is a proper face F of P such that $N \subset \text{ext } F$, then the connectedness of Γ follows from Theorem 15.5. If no such face F exists, then—assuming that $P \subset \mathbb{R}^d$—any hyperplane containing N must intersect int P. Choose such a hyperplane H containing at least one more vertex x_0 of P. Let K_1 and K_2 be the two closed halfspaces bounded by H, and let $P_1 := K_1 \cap P$ and $P_2 := K_2 \cap P$. Then P_1 and P_2 are d-polytopes, cf. Theorem 9.2. Moreover, the set $F := H \cap P$ is a facet of both. The vertices of F are the vertices of P in H and the 1-point intersections of H and edges of P crossing H, cf. Theorem 11.1(b), (d). Let

$$M := N \cup (\text{ext } F\backslash\text{ext } P).$$

Let Γ_1 denote the subgraph of $\mathscr{G}(P_1)$ spanned by (ext $P_1)\backslash M$, and let Γ_2 denote the subgraph of $\mathscr{G}(P_2)$ spanned by (ext $P_2)\backslash M$. Theorem 15.5 states that Γ_1 and Γ_2 are connected. Now, let x be any vertex of P not in N. Then x is a vertex of Γ_1 or Γ_2 (or both); assume that x is a vertex of Γ_1. Then by the connectedness of Γ_1 there is a path in Γ_1 joining x and x_0. This is a path in $\mathscr{G}(P)$ not entering N. Hence, any two vertices of P not in N can be joined by a path in $\mathscr{G}(P)$ not entering N via the vertex x_0, showing that the subgraph of $\mathscr{G}(P)$ spanned by (ext $P)\backslash N$ is connected. □

By a *facet system* in a polytope P we mean a non-empty set $\mathscr{S}$ of facets of P. Each $\mathscr{G}(F)$, $F \in \mathscr{S}$, is then a subgraph of $\mathscr{G}(P)$. The union of the subgraphs $\mathscr{G}(F)$, $F \in \mathscr{S}$, is denoted by $\mathscr{G}(\mathscr{S})$. We shall say that a facet system $\mathscr{S}$ is *connected* if $\mathscr{G}(\mathscr{S})$ is a connected graph.

Theorem 15.7. *Let $\mathscr{S}$ be a connected facet system in a simple d-polytope P, where $d \geq 2$. Then $\mathscr{G}(\mathscr{S})$ is a $(d - 1)$-connected graph.*

Proof. We prove the statement by induction on the number n of members of $\mathscr{S}$. For $n = 1$, the statement follows immediately from Theorem 15.6. For $n \geq 2$, we may number the members $F_1, \ldots, F_n$ of $\mathscr{S}$ in such a manner that the subsystem $\mathscr{S}'$ formed by $F_1, \ldots, F_{n-1}$ is also connected. (Take F_1 arbitrary, use the connectedness of $\mathscr{S}$ to find F_2 such that $\{F_1, F_2\}$ is connected, use the connectedness of $\mathscr{S}$ to find F_3 such that $\{F_1, F_2, F_3\}$ is connected, etc.). Then use the induction hypothesis to deduce that $\mathscr{G}(\mathscr{S}')$ is $(d - 1)$-connected. Now, by the connectedness of $\mathscr{S}$ there is an F_j with $j \leq n - 1$ such that $F_j \cap F_n \neq \emptyset$. It then follows from Theorem 12.14 that $F_j \cap F_n$ is a $(d - 2)$-face of P. Therefore, F_j and F_n have at least $d - 1$ vertices in common. Hence, the graphs $\mathscr{G}(\mathscr{S}')$ and $\mathscr{G}(F_n)$ have at least $d - 1$ vertices in common. Since both graphs are $(d - 1)$-connected, it follows that their union, i.e. $\mathscr{G}(\mathscr{S})$, is $(d - 1)$-connected, cf. Appendix 2, Theorem A2.2. □

EXERCISES

15.1. A *semi-shelling* of a d-polytope P is a numbering $F_1, F_2, \ldots, F_k$ of the facets of P such that for $i = 2, \ldots, k$, the set

$$F_i \cap \bigcup_{j=1}^{i-1} F_j \tag{1}$$

is a non-empty union of $(d-2)$-faces of F_i. Show by a duality argument that every d-polytope admits semi-shellings.

Verify that any facet can be taken as F_1. Verify that the facets containing a given face can be taken to precede all the remaining facets.

(A semi-shelling is a *shelling* if, in addition, for $i = 2, \ldots, k-1$ the set (1) is homeomorphic to a $(d-2)$-ball. If P is simplicial, then this condition is automatically fulfilled.)

15.2. A graph is said to be *planar* if, loosely speaking, it can be drawn in the plane with non-intersecting (not necessarily rectilinear) edges. Show that the graph $\mathscr{G}(P)$ of any 3-polytope P is planar. (Along with Theorem 15.6, this proves the easy part of *Steinitz's Theorem*: A graph Γ is (isomorphic to) the graph $\mathscr{G}(P)$ of a 3-polytope P if and only if it is planar and 3-connected.)

CHAPTER 3

Combinatorial Theory of Convex Polytopes

§16. Euler's Relation

At the beginning of Section 10 it was indicated that the combinatorial theory of convex polytopes may be described as the study of their face-lattices. When it comes to reality, however, this description is too ambitious. Instead, we shall describe the combinatorial theory as the study of f-vectors. For $d \geq 1$, the f-vector of a d-polytope P is the d-tuple

$$f(P) = (f_0(P), f_1(P), \ldots, f_{d-1}(P)).$$

where $f_j(P)$ denotes the number of j-faces of P, cf. Section 10. Equivalent polytopes have the same f-vector, but the converse is not true in general.

It may be said that the basic problem is as follows: Which d-tuples of positive integers are the f-vectors of d-polytopes? Denoting by $\mathscr{P}^d$ the set of all d-polytopes and by $f(\mathscr{P}^d)$ the set of all f-vectors of d-polytopes, the problem amounts to determining the subset $f(\mathscr{P}^d)$ of $\mathbb{R}^d$. This problem has only been solved completely for $d \leq 3$, the cases $d = 1, 2$ being trivial.

In this section we shall determine the affine hull $\operatorname{aff} f(\mathscr{P}^d)$ of the set $f(\mathscr{P}^d)$. This partial solution to the basic problem is a main general result in the area.

We first prove that there is a linear relation which is satisfied by the numbers $f_j(P), j = 0, \ldots, d-1$, for any d-polytope P. For technical reasons, we prefer to include the numbers $f_{-1}(P) = 1$ and $f_d(P) = 1$. The relation is known as *Euler's Relation*:

Theorem 16.1. *For any d-polytope P one has*

$$\sum_{j=-1}^{d} (-1)^j f_j(P) = 0.$$

Note that Euler's Relation may also be written as

$$\sum_{j=0}^{d}(-1)^j f_j(P) = 1,$$

or

$$\sum_{j=0}^{d-1}(-1)^j f_j(P) = 1 - (-1)^d$$
$$= 1 + (-1)^{d-1}.$$

Since $f_j(P) = 0$ when $j > d = \dim P$, we may also write

$$\sum_{j \geq -1}(-1)^j f_j(P) = 0,$$

thus avoiding reference to the dimension of P.

PROOF. We use induction on d. For $d = 0, 1$ there is nothing to prove and for $d = 2$ the statement is obvious. So, let d be at least 3, assume that the statement is valid for all polytopes of dimension $\leq d - 1$, and let P be a d-polytope. Assuming that $P \subset \mathbb{R}^d$, we choose an admissible vector w for P, cf. Theorem 15.1. Let $x_1, \ldots, x_n$ be the vertices of P, numbered such that

$$\langle x_i, w\rangle < \langle x_{i+1}, w\rangle, \qquad i = 1, \ldots, n-1.$$

Calling w the down direction as we did in Section 15, this means that x_{i+1} is below x_i. Let

$$\alpha_{2i-1} := \langle x_i, w\rangle, \qquad i = 1, \ldots, n.$$

Noting that $\alpha_{2i-1} < \alpha_{2i+1}$, we next choose α_{2i} such that

$$\alpha_{2i-1} < \alpha_{2i} < \alpha_{2i+1}, \qquad i = 1, \ldots, n-1,$$

and define

$$H_k := H(w, \alpha_k), \qquad k = 1, \ldots, 2n-1.$$

Then the H_k's form a collection of parallel hyperplanes with H_{k+1} below H_k such that the H_k's with odd values of k pass through the vertices of P. Let

$$P_k := H_k \cap P, \qquad k = 1, \ldots, 2n-1.$$

Then P_k is a $(d-1)$-polytope for $k = 2, \ldots, 2n-2$, whereas $P_1 = \{x_1\}$ and $P_{2n-1} = \{x_n\}$. By the induction hypothesis, Euler's Relation is valid for the polytopes P_k, whence

$$\sum_{j \geq -1}(-1)^j f_j(P_k) = 0, \qquad k = 1, \ldots, 2n-1.$$

Multiplying by $(-1)^{k+1}$ and adding, we get

$$\sum_{k=1}^{2n-1}(-1)^{k+1}\sum_{j \geq -1}(-1)^j f_j(P_k) = 0,$$

which we may rewrite as

$$\sum_{j \geq -1} (-1)^{j+1} \sum_{k=1}^{2n-1} (-1)^k f_j(P_k) = 0. \tag{1}$$

We shall prove that

$$\sum_{k=1}^{2n-1} (-1)^k f_j(P_k) = \begin{cases} -1, & j = -1; \\ f_1(P) - f_0(P), & j = 0; \\ f_{j+1}(P), & j = 1, \ldots, d-1. \end{cases} \tag{2}$$

Combining (1) and (2), we obtain the desired relation.

To prove (2) for $j = -1$, note that $f_{-1}(P_k) = 1$ for all k, whence the left-hand side is an alternating sum of the form $-1 + 1 - \cdots + 1 - 1$, which has the value -1.

To prove (2) for $j \geq 0$, we define for $F \in \mathscr{F}(P)$ and $k = 1, \ldots, 2n-1$,

$$\psi(F, k) := \begin{cases} 1 & \text{if } H_k \cap \text{ri } F \neq \emptyset; \\ 0 & \text{if } H_k \cap \text{ri } F = \emptyset. \end{cases}$$

Furthermore, we denote by $\mathscr{F}_j(P)$ the set of j-faces of P.

Let us first consider the case $1 \leq j \leq d-1$. Here a j-face of P has at least two vertices, and since each H_k contains at most one vertex of P, it follows that no j-face of P_k is a face of P. Theorem 11.1(d) then shows that for each j-face F' of P_k there is a unique face F of P such that $F' = H_k \cap F$, and this face F is a $(j+1)$-face. Under these circumstances, it is clear that $H_k \cap \text{ri } F \neq \emptyset$. Conversely, if F is a $(j+1)$-face of P with $H_k \cap \text{ri } F \neq \emptyset$, then $F' := H_k \cap F$ is a j-face of P_k, cf. Theorem 11.1(b). In conclusion, for fixed j and k, the mapping

$$F \mapsto F' := H_k \cap F$$

is a one-to-one mapping from the set of $(j+1)$-faces F of P with $\psi(F, k) = 1$ onto the set of j-faces of P_k. Therefore,

$$f_j(P_k) = \sum_{F \in \mathscr{F}_{j+1}(P)} \psi(F, k). \tag{3}$$

Having established (3), we may rewrite the left-hand side of (2):

$$\begin{aligned} \sum_{k=1}^{2n-1} (-1)^k f_j(P_k) &= \sum_{k=1}^{2n-1} (-1)^k \sum_{F \in \mathscr{F}_{j+1}(P)} \psi(F, k) \\ &= \sum_{F \in \mathscr{F}_{j+1}(P)} \sum_{k=1}^{2n-1} (-1)^k \psi(F, k). \end{aligned} \tag{4}$$

Now, let us consider a fixed $(j+1)$-face F of P. Let x_{i_1} be the top vertex of F, and let x_{i_2} be the bottom vertex of F. Then the values of k such that $\psi(F, k) = 1$

are the values $k = 2i_1, 2i_1 + 1, \ldots, 2i_2 - 3, 2i_2 - 2$. Here the number of even values is 1 larger than the number of odd values, whence

$$\sum_{k=1}^{2n-1} (-1)^k \psi(F, k) = 1. \tag{5}$$

Combining (4) and (5) we then get

$$\sum_{k=1}^{2n-1} (-1)^k f_j(P_k) = \sum_{F \in \mathscr{F}_{j+1}(P)} 1$$
$$= f_{j+1}(P),$$

proving (2) for $1 \leq j \leq d - 1$.

The case $j = 0$ requires a little more care. For k even, H_k contains no vertex of P. It then follows that for k even, no 0-face of P_k is a face of P. We can then argue as above and we obtain

$$f_0(P_k) = \sum_{F \in \mathscr{F}_1(P)} \psi(F, k), \qquad k \text{ even}. \tag{6}$$

For k odd, the situation is slightly different. In this case, H_k contains one vertex of P which is then also a vertex of P_k. For the remaining vertices of P_k we can next argue as above, thus obtaining

$$f_0(P_k) = \sum_{F \in \mathscr{F}_1(P)} \psi(F, k) + 1, \qquad k \text{ odd}. \tag{7}$$

Using (6) and (7), the left-hand side of (2) may be rewritten as

$$\sum_{k=1}^{2n-1} (-1)^k f_0(P_k) = \sum_{k=1}^{2n-1} (-1)^k \sum_{F \in \mathscr{F}_1(P)} \psi(F, k) + \sum_{\substack{k=1 \\ k \text{ odd}}}^{2n-1} (-1)^k$$
$$= \sum_{F \in \mathscr{F}_1(P)} \sum_{k=1}^{2n-1} (-1)^k \psi(F, k) - f_0(P), \tag{8}$$

where we have used the fact that the number of odd values of k equals $f_0(P)$. We next argue as in the case $1 \leq j \leq d - 1$ to obtain

$$\sum_{k=1}^{2n-1} (-1)^k \psi(F, k) = 1 \tag{9}$$

for any $F \in \mathscr{F}_1(P)$. Combining (8) and (9) we then get

$$\sum_{k=1}^{2n-1} (-1)^k f_0(P_k) = \sum_{F \in \mathscr{F}_1(P)} 1 - f_0(P)$$
$$= f_1(P) - f_0(P),$$

proving (2) for $j = 0$. This completes the proof. □

Now, for $d \geq 1$ let

$$\varepsilon := (1, -1, \ldots, (-1)^{d-1}) \in \mathbb{R}^d.$$

Then $H(\varepsilon, 1-(-1)^d)$ is a hyperplane in $\mathbb{R}^d$ which we shall call the *Euler Hyperplane*. Note that $o \in H(\varepsilon, 1-(-1)^d)$ if and only if d is even. Theorem 16.1 shows that $f(\mathscr{P}^d)$ is contained in the Euler Hyperplane. We shall prove:

Theorem 16.2. *The Euler Hyperplane $H(\varepsilon, 1-(-1)^d)$ is the only hyperplane in $\mathbb{R}^d$ which contains $f(\mathscr{P}^d)$.*

Proof. We use induction on d. For $d = 1, 2$, the statement is obvious. So, let d be at least 3 and assume that the statement holds for all dimensions $\leq d-1$. Let $H = H(y, \alpha)$ be any hyperplane in $\mathbb{R}^d$ such that $f(\mathscr{P}^d) \subset H(y, \alpha)$. We shall prove that $H(y, \alpha) = H(\varepsilon, 1-(-1)^d)$ by showing that there is a real $c \neq 0$ such that $y = c\varepsilon$ and $\alpha = c(1-(-1)^d)$.

Let Q be any $(d-1)$-polytope, and let Q' be an equivalent of Q in $\mathbb{R}^d$. Let P_1 be a d-pyramid in $\mathbb{R}^d$ with Q' as a basis, and let P_2 be a d-bipyramid with Q' as a basis. We can express $f(P_1)$ and $f(P_2)$ in terms of $f(Q') = f(Q)$. In fact, it follows from Theorem 7.7 that

$$f(P_1) = (f_0(Q) + 1, f_1(Q) + f_0(Q), \ldots, f_{d-2}(Q) + f_{d-3}(Q), 1 + f_{d-2}(Q)),$$

and it follows from Theorem 7.8 that

$$f(P_2) = (f_0(Q) + 2, f_1(Q) + 2f_0(Q), \ldots, f_{d-2}(Q) + 2f_{d-3}(Q), 2f_{d-2}(Q)).$$

Now, writing

$$y = (\alpha_1, \ldots, \alpha_d),$$

we have

$$\alpha_1(f_0(Q) + 1) + \alpha_2(f_1(Q) + f_0(Q)) + \cdots \\ \cdots + \alpha_{d-1}(f_{d-2}(Q) + f_{d-3}(Q)) + \alpha_d(1 + f_{d-2}(Q)) = \alpha \quad (10)$$

since $f(P_1) \in H(y, \alpha)$, and we have

$$\alpha_1(f_0(Q) + 2) + \alpha_2(f_1(Q) + 2f_0(Q)) + \cdots \\ \cdots + \alpha_{d-1}(f_{d-2}(Q) + 2f_{d-3}(Q)) + \alpha_d 2f_{d-2}(Q) = \alpha \quad (11)$$

since $f(P_2) \in H(y, \alpha)$. Subtraction of (10) from (11) yields

$$\alpha_1 + \alpha_2 f_0(Q) + \cdots + \alpha_{d-1} f_{d-3}(Q) + \alpha_d(f_{d-2}(Q) - 1) = 0. \quad (12)$$

We reqrite (12) as

$$\alpha_2 f_0(Q) + \alpha_3 f_1(Q) + \cdots + \alpha_{d-1} f_{d-3}(Q) + \alpha_d f_{d-2}(Q) = \alpha_d - \alpha_1. \quad (13)$$

Letting

$$z := (\alpha_2, \ldots, \alpha_d),$$

it follows from (13) that

$$f(Q) \in H(z, \alpha_d - \alpha_1). \quad (14)$$

Since $y \neq o$ by assumption, we have $\alpha_j \neq 0$ for at least one $j = 1, \ldots, d$. But then also $\alpha_j \neq 0$ for at least one $j = 2, \ldots, d$, implying that $H(z, \alpha_d - \alpha_1)$ is a hyperplane in $\mathbb{R}^{d-1}$. As Q is arbitrary, we see that $H(z, \alpha_d - \alpha_1)$ is a hyperplane in $\mathbb{R}^{d-1}$ containing $f(\mathscr{P}^{d-1})$, cf. (14). By the induction hypothesis, this implies that

$$(\alpha_2, \ldots, \alpha_d) = \gamma(1, -1, \ldots, (-1)^{d-2})$$

and

$$\alpha_d - \alpha_1 = \gamma(1 - (-1)^{d-1})$$

for a suitable $\gamma \neq 0$. Taking $c := -\gamma$, we then see that $y = c\varepsilon$ and $\alpha = c(1 - (-1)^d)$, as desired. □

An immediate consequence of Theorems 16.1 and 16.2 is the following main result:

Corollary 16.3. *The affine hull* $\operatorname{aff} f(\mathscr{P}^d)$ *of* $f(\mathscr{P}^d)$ *is the Euler Hyperplane* $H(\varepsilon, 1 - (-1)^d)$.

We conclude this section with a variant of Euler's Relation. For faces F_1 and F_2 of a d-polytope P with $F_1 \subset F_2$ we shall write $f_j(F_2/F_1)$ for the number of j-faces F of P such that $F_1 \subset F \subset F_2$. Note that $f_j(F_2/F_1) = 0$ for $-1 \leq j < \dim F_1$ and for $\dim F_2 < j$. We shall establish a linear relation between the numbers $f_j(F_2/F_1)$ when $F_1 \subsetneqq F_2$.

Theorem 16.4. *Let F_1 and F_2 be faces of a polytope P with $F_1 \subsetneqq F_2$. Then*

$$\sum_{j \geq -1} (-1)^j f_j(F_2/F_1) = 0.$$

PROOF. We know by Theorem 11.4 that there is a polytope Q with

$$\dim Q = \dim F_2 - 1 - \dim F_1$$

such that the lattice $(\mathscr{F}(F_2/F_1), \subset)$ is isomorphic to the face-lattice $(\mathscr{F}(Q), \subset)$, and, moreover,

$$\dim G = \dim F - 1 - \dim F_1$$

when $F_1 \subset F \subset F_2$ and G is the face of Q corresponding to F. In particular,

$$f_j(F_2/F_1) = f_k(Q) \tag{15}$$

when

$$k = j - 1 - \dim F_1.$$

Euler's Relation for Q may be written as

$$\sum_{k \geq -1} (-1)^k f_k(Q) = 0.$$

Using (15), we next obtain

$$\sum_{j \geq -1} (-1)^{j-1-\dim F_1} f_j(F_2/F_1) = 0,$$

which is clearly equivalent to the desired relation. □

Of course, in Theorem 16.4 we recover Euler's Relation by taking $F_1 = \emptyset$ and $F_2 = P$.

§17. The Dehn–Sommerville Relations

In the preceding section we showed that the affine hull of $f(\mathscr{P}^d)$ has dimension $d-1$. Denoting by $\mathscr{P}^d_\sigma$ the set of all simple d-polytopes, and by $f(\mathscr{P}^d_\sigma)$ the set of all f-vectors of simple d-polytopes, we shall prove that the dimension of $\operatorname{aff} f(\mathscr{P}^d_\sigma)$ is only $\lfloor d/2 \rfloor$. Moreover, we shall find "representations" of $\operatorname{aff} f(\mathscr{P}^d_\sigma)$.

(There is no standard notation for the set of simple d-polytopes. Our notation $\mathscr{P}^d_\sigma$ is inspired by the standard notation $\mathscr{P}^d_s$ for the set of simplicial d-polytopes.)

We first exhibit a set of linear relations which are satisfied by the f-vectors of all simple d-polytopes. These relations are known as the *Dehn–Sommerville Relations*:

Theorem 17.1. *For any simple d-polytope P we have*

$$\sum_{j=0}^{d} (-1)^j \binom{d-j}{d-i} f_j(P) = f_i(P)$$

for $i = 0, \ldots, d$.

Note that, effectively, we only sum from $j = 0$ to $j = i$. For $i = d$ we get Euler's Relation. For $i = 0$ we get the trivial relation $f_0(P) = f_0(P)$.

For $d \leq 2$, everything is trivial. For $d = 3$, the relations are

$$\begin{aligned} f_0(P) \qquad\qquad\qquad\qquad &= f_0(P),\\ 3f_0(P) - \ f_1(P) \qquad\qquad\quad &= f_1(P),\\ 3f_0(P) - 2f_1(P) + f_2(P) \qquad &= f_2(P),\\ f_0(P) - \ f_1(P) + f_2(P) - 1 &= 1. \end{aligned}$$

These four relations are equivalent to the following two:

$$\begin{aligned} 3f_0(P) - 2f_1(P) \qquad\quad &= 0,\\ f_0(P) - \ f_1(P) + f_2(P) &= 2. \end{aligned}$$

It is interesting to note that if one of the three numbers $f_0(P)$, $f_1(P)$, and $f_2(P)$ is known, then the remaining two are determined by the Dehn–Sommerville Relations. In particular, we can express $f_0(P)$ and $f_1(P)$ by $f_2(P)$:

$$f_0(P) = 2f_2(P) - 4, \qquad f_1(P) = 3f_2(P) - 6.$$

(See Theorem 17.6 and Corollary 17.7 below for a d-dimensional version.)

PROOF. For any non-empty face F of P, Euler's Relation states that

$$\sum_{j \geq -1} (-1)^j f_j(F) = 0. \tag{1}$$

Using the notation $\mathscr{F}_i(P)$ for the set of i-faces of P as we did in the proof of Euler's Relation, it follows immediately from (1) that for $i = 0, \ldots, d$ we have

$$\sum_{F \in \mathscr{F}_i(P)} \sum_{j \geq -1} (-1)^j f_j(F) = 0,$$

or, equivalently,

$$\sum_{j \geq -1} (-1)^j \sum_{F \in \mathscr{F}_i(P)} f_j(F) = 0. \tag{2}$$

The value of the sum

$$\sum_{F \in \mathscr{F}_i(P)} f_j(F)$$

is the number of pairs (G, F) of faces of P such that $\dim G = j$, $\dim F = i$ and $G \subset F$. Therefore,

$$\sum_{F \in \mathscr{F}_i(P)} f_j(F) = \sum_{G \in \mathscr{F}_j(P)} f_i(P/G). \tag{3}$$

For $\dim G = j \geq 0$, the number $f_i(P/G)$ was determined in Theorem 12.16; in fact, we showed that

$$f_i(P/G) = \binom{d-j}{d-i}. \tag{4}$$

Note that for $i < j \leq d$, both sides of (4) are 0, and so (4) is valid for $0 \leq j \leq d$. For $\dim G = j = -1$, it is clear that

$$f_i(P/G) = f_i(P). \tag{5}$$

Combining now (2), (3), (4), and (5), we get the desired relation. □

As mentioned in the beginning, we aim to show that the dimension of $\operatorname{aff} f(\mathscr{P}_\sigma^d)$ is $\lfloor d/2 \rfloor$. It will follow from Theorem 17.1 that the dimension is at most $\lfloor d/2 \rfloor$. To see that it is at least $\lfloor d/2 \rfloor$, we need the lemma below. To ease the notation, we shall write

$$m := \lfloor (d-1)/2 \rfloor, \qquad n := \lfloor d/2 \rfloor.$$

Note that $d = m + n + 1$.

Lemma 17.2. *For $i = 0, \ldots, n$, let P_i be a cyclic polytope of type $C(p + i, d)$ for some fixed $p \geq d + 1$. Then the f-vectors $f(P_i)$, $i = 0, \ldots, n$, form an affinely independent family in $\mathbb{R}^d$.*

PROOF. The f-vector of P_i has the form

$$f(P_i) = \left(\binom{p+i}{1}, \ldots, \binom{p+i}{n}, f_n(P_i), \ldots, f_{d-1}(P_i)\right),$$

cf. Theorem 13.5 and Corollary 13.8. Saying that the $f(P_i)$'s are affinely independent is equivalent to saying that the $\bar{f}(P_i)$'s defined by

$$\bar{f}(P_i) := \left(1, \binom{p+i}{1}, \ldots, \binom{p+i}{n}, f_n(P_i), \ldots, f_{d-1}(P_i)\right),$$

are linearly independent. In terms of matrices, this is equivalent to saying that the $(n + 1) \times (d + 1)$ matrix whose rows are formed by the $\bar{f}(P_i)$'s has rank $n + 1$. Consider the submatrix A formed by the first $n + 1$ columns, i.e.

$$A := \left(\binom{p+i}{j}\right)_{i=0,\ldots,n;\, j=0,\ldots,n}.$$

We shall complete the proof by showing that A is invertible. Let

$$B := \left(\binom{-p+j}{n-i}\right)_{i=0,\ldots,n;\, j=0,\ldots,n}$$

and let

$$C := AB = (c_{ij})_{i=0,\ldots,n;\, j=0,\ldots,n}.$$

Then

$$c_{ij} = \sum_{k=0}^{n} \binom{p+i}{k}\binom{-p+j}{n-k} = \binom{i+j}{n},$$

cf. Appendix 3, (7). Hence, C has only 1's in the "skew diagonal" (i.e. the positions (i, j) with $i + j = n$) and only 0's above. Therefore,

$$|\det C| = 1,$$

implying that A is invertible. □

It is easy to prove by induction on n that the matrix A in the proof above has determinant 1. Hence, one can prove that A is invertible without referring to Appendix 3, (7), if desired.

We shall next prove:

Theorem 17.3. *The affine hull* aff $f(\mathscr{P}_\sigma^d)$ *of* $f(\mathscr{P}_\sigma^d)$ *has dimension* $\lfloor d/2 \rfloor$.

PROOF. Consider the system of $d + 1$ (homogeneous) linear equations

$$\sum_{j=0}^{d} (-1)^j \binom{d-j}{d-i} x_j = x_i, \qquad i = 0, \ldots, d, \tag{6}$$

with unknowns $x_0, \ldots, x_{d-1}, x_d$. Assigning the value 1 to x_d we obtain a system of $d + 1$ linear equations with d unknowns $x_0, \ldots, x_{d-1}$. (Note that for d odd, the equation corresponding to $i = d$ is inhomogeneous.) Theorem 17.1 tells that for each polytope $P \in \mathscr{P}_\sigma^d$,

$$(x_0, \ldots, x_{d-1}) = (f_0(P), \ldots, f_{d-1}(P))$$

is a solution. In other words, denoting the set of solutions by S, we have $f(\mathscr{P}_\sigma^d) \subset S$, whence

$$\operatorname{aff} f(\mathscr{P}_\sigma^d) \subset \operatorname{aff} S = S,$$

and so

$$\dim(\operatorname{aff} f(\mathscr{P}_\sigma^d)) \leq \dim S.$$

Now, it is easy to see that the equations (6) corresponding to odd values of i are independent. Since the number of odd values of i is $m + 1$, it follows that

$$\begin{aligned}\dim S &\leq d - (m + 1)\\ &= n,\end{aligned}$$

implying that

$$\dim(\operatorname{aff} f(\mathscr{P}_\sigma^d)) \leq n.$$

To prove the converse, we apply Lemma 17.2. Let P_i be as described there, and let Q_i be a dual of P_i, $i = 0, \ldots, n$. Then $f(Q_i) \in f(\mathscr{P}_\sigma^d)$ by Theorems 13.5 and 12.10. Moreover, since the $f(P_i)$'s are affinely independent, cf. Lemma 17.2, it follows that the $f(Q_i)$'s are affinely independent. The number of $f(Q_i)$'s is $n + 1$, and therefore the dimension of $\operatorname{aff} f(\mathscr{P}_\sigma^d)$ is at least n. This completes the proof. □

For $i = 0, \ldots, d$, let H_i denote the set of points $(x_0, \ldots, x_{d-1}) \in \mathbb{R}^d$ such that

$$\sum_{j=0}^{d} (-1)^j \binom{d-j}{d-i} x_j = x_i,$$

where, as in the proof above, it is understood that $x_d = 1$. Then $H_0 = \mathbb{R}^d$, and for $i \geq 1$, each H_i is a hyperplane in $\mathbb{R}^d$. During the proof above, we showed that

$$n = \dim \bigcap_{\substack{i=0\\ i \text{ odd}}}^{d} H_i \geq \dim \bigcap_{i=0}^{d} H_i \geq \dim(\operatorname{aff} f(\mathscr{P}_\sigma^d)) \geq n.$$

We also have

$$\operatorname{aff} f(\mathscr{P}_\sigma^d) \subset \bigcap_{i=0}^{d} H_i \subset \bigcap_{\substack{i=0\\ i \text{ odd}}}^{d} H_i,$$

and we can therefore conclude that

$$\operatorname{aff} f(\mathscr{P}_\sigma^d) = \bigcap_{i=0}^{d} H_i = \bigcap_{\substack{i=0 \\ i \text{ odd}}}^{d} H_i.$$

Thus, we have obtained "representations" of $\operatorname{aff} f(\mathscr{P}_\sigma^d)$ as intersections of families of hyperplanes. (Since $H_0 = \mathbb{R}^d$, this set can be omitted.) Actually, the "representation" of $\operatorname{aff} f(\mathscr{P}_\sigma^d)$ as the intersection of the odd-numbered H_i's is "minimal" in the sense that it includes the smallest possible number of hyperplanes. In the following, we shall establish other such "representations." We shall, however, prefer to formulate the results in terms of linear equations, rather than using a geometric terminology.

A system of linear equations with d unknowns $x_0, \ldots, x_{d-1}$ will be called a *Dehn–Sommerville System* for the simple d-polytopes if its set of solutions is precisely $\operatorname{aff} f(\mathscr{P}_\sigma^d)$. A Dehn–Sommerville System containing a minimal number of equations is said to be *minimal*. It follows from Theorem 17.3 that this minimal number is

$$d - \lfloor d/2 \rfloor = \lfloor (d+1)/2 \rfloor.$$

Any subsystem of a Dehn–Sommerville System which is formed by $\lfloor (d+1)/2 \rfloor$ independent equations is necessarily again a Dehn–Sommerville System (and hence minimal).

It follows immediately from the remarks above that we have:

Theorem 17.4. *The equations*

$$\sum_{j=0}^{d} (-1)^j \binom{d-j}{d-i} x_j = x_i, \qquad i = 0, \ldots, d,$$

where $x_d = 1$, form a Dehn–Sommerville System. The equations corresponding to odd values of i form a minimal Dehn–Sommerville System.

In dealing with Dehn–Sommerville Systems it is convenient to use matrix notation. We shall write

$$x := \begin{pmatrix} x_0 \\ \vdots \\ x_{d-1} \\ x_d \end{pmatrix},$$

where it is always understood that $x_d = 1$. If we let A be the $(d+1) \times (d+1)$ matrix defined by

$$A := \left((-1)^j \binom{d-j}{d-i} \right)_{i=0,\ldots,d;\, j=0,\ldots,d}$$

then we may write the Dehn–Sommerville System of Theorem 17.4 as

$$Ax = x. \tag{7}$$

Now, once we know this Dehn–Sommerville System, it is easy to produce new ones: any system obtained from (7) by multiplying from the left on both sides by an invertible matrix will again be a Dehn–Sommerville System. We shall apply this procedure below.

Theorem 17.5. *The equations*

$$\sum_{j=0}^{d}(-1)^j\binom{j}{i}x_j = \sum_{j=0}^{d}(-1)^{d+j}\binom{j}{d-i}x_j, \qquad i = 0, \ldots, d,$$

where $x_d = 1$, *form a Dehn–Sommerville System. The equations corresponding to the values* $i = 0, \ldots, m$ *form a minimal Dehn–Sommerville System.*

Proof. Let B be the $(d+1) \times (d+1)$ matrix defined by

$$B := \left((-1)^{i+j}\binom{j}{i}\right)_{i=0,\ldots,d;\, j=0,\ldots,d}.$$

Note that B is invertible, cf. Appendix 3, (11). Since $x = Ax$ is a Dehn–Sommerville System, it follows that

$$Bx = BAx \tag{8}$$

is also a Dehn–Sommerville System. Except for the factor $(-1)^i$, the ith entry on the left-hand side of (8) is the left-hand side of the ith equation in the theorem. In order to evaluate the right-hand side of (8), we first calculate the element of BA in the ith row and jth column. This element is

$$\begin{aligned}\sum_{k=0}^{d}(-1)^{i+k}\binom{k}{i}(-1)^j\binom{d-j}{d-k} &= (-1)^{i+j}\sum_{k=0}^{d}(-1)^k\binom{k}{i}\binom{d-j}{d-k}\\ &= (-1)^{i+j}(-1)^d\binom{j}{d-i},\end{aligned}$$

where we have used Appendix 3, (12). Then the ith entry on the right-hand side of (8) becomes

$$\sum_{j=0}^{d}(-1)^{i+j+d}\binom{j}{d-i}x_j = (-1)^i\sum_{j=0}^{d}(-1)^{d+j}\binom{j}{d-i}x_j,$$

which — except for the factor $(-1)^i$—is the right-hand side of the ith equation. This completes the proof of the first statement.

To see that the first $m+1$ equations form a minimal Dehn–Sommerville System, if suffices to show that they are independent. We have proved above that—except for the factor $(-1)^i$—the equations of the theorem may be written as

$$Bx = BAx.$$

We rewrite this as

$$(B - BA)x = 0,$$

where 0 on the right-hand side denotes the $(d + 1) \times 1$ zero matrix. Now, note that the calculation above shows that the element of BA in the ith row and jth column is 0 for $i, j \leq m$. In other words, the $(m + 1) \times (m + 1)$ submatrix of $B - BA$ in the upper-left corner is the matrix

$$B_0 := \left((-1)^{i+j}\binom{j}{i}\right)_{i=0,\ldots,m;\, j=0,\ldots,m}$$

This matrix, however, is invertible, cf. Appendix 3, (11), and therefore the first $m + 1$ equations are independent. □

Theorem 17.6. *The equations*

$$x_i = \sum_{j=0}^{n}(-1)^j\binom{m+1+j}{i}\binom{m-i+j}{m-i}x_{m+1+j}$$
$$+ (-1)^n\sum_{j=0}^{n}(-1)^j\left(\sum_{k=0}^{m}(-1)^k\binom{k}{i}\binom{m+1+j}{d-k}\right)x_{m+1+j}, \qquad i = 0, \ldots, m,$$

where $x_d = 1$, *form a minimal Dehn–Sommerville System.*

Proof. Let B_2 be the $(m + 1) \times (d + 1)$ matrix defined by

$$B_2 := \left((-1)^{i+j}\binom{j}{i}\right)_{i=0,\ldots,m;\, j=0,\ldots,d}$$

and let C be the $(m + 1) \times (d + 1)$ matrix defined by

$$C := \left((-1)^{d+i+j}\binom{j}{d-i}\right)_{i=0,\ldots,m;\, j=0,\ldots,d}.$$

Then the minimal Dehn–Sommerville System of Theorem 17.5 may be written as

$$B_2 x = Cx. \tag{9}$$

Let B_0 and B_1 be the submatrices of B_2 formed by the first $m + 1$ and the last $(d + 1) - (m + 1) = n + 1$ columns of B_2, respectively. (Then B_0 denotes the same matrix as in the proof of Theorem 17.6.) In a similar way, let C_0 and C_1 denote the submatrices of C formed by the first $m + 1$ and the last $n + 1$ columns of C, respectively. Then we may rewrite (9) as

$$(B_0 - C_0)\begin{pmatrix} x_0 \\ \vdots \\ x_m \end{pmatrix} = (C_1 - B_1)\begin{pmatrix} x_{m+1} \\ \vdots \\ x_d \end{pmatrix}. \tag{10}$$

Now, note that C_0 is a zero matrix, whence $B_0 - C_0 = B_0$. Let

$$D_0 := \left(\binom{j}{i}\right)_{i=0,\ldots,m;\, j=0,\ldots,m}.$$

Then D_0 is in fact the inverse of B_0, cf. Appendix 3, (11). Therefore, multiplication by D_0 in (10) gives the new minimal Dehn–Sommerville System

$$\begin{pmatrix} x_0 \\ \vdots \\ x_m \end{pmatrix} = D_0(C_1 - B_1)\begin{pmatrix} x_{m+1} \\ \vdots \\ x_d \end{pmatrix}.$$

To see that this is the system in the theorem, we calculate the elements of $D_0(C_1 - B_1)$. Note first that

$$C_1 - B_1 = \left((-1)^{n+i+j}\binom{m+1+j}{d-i} + (-1)^{m+i+j}\binom{m+1+j}{i}\right)_{i=0,\ldots,m;\, j=0,\ldots,n}.$$

Then the element of $D_0(C_1 - B_1)$ in the ith row and jth column is

$$\begin{aligned}
&\sum_{k=0}^{m}\binom{k}{i}\left((-1)^{n+k+j}\binom{m+1+j}{d-k} + (-1)^{m+k+j}\binom{m+1+j}{k}\right) \\
&= \sum_{k=0}^{m}(-1)^{m+k+j}\binom{k}{i}\binom{m+1+j}{k} + \sum_{k=0}^{m}(-1)^{n+k+j}\binom{k}{i}\binom{m+1+j}{d-k} \\
&= (-1)^{j}\binom{m+1+j}{i}\binom{m-i+j}{m-i} + \sum_{k=0}^{m}(-1)^{n+k+j}\binom{k}{i}\binom{m+1+j}{d-k},
\end{aligned}$$

cf. Appendix 3, (10). From this the statement follows immediately. □

We note an interesting corollary of Theorem 17.6:

Corollary 17.7. *Let P be a simple d-polytope. Then the numbers $f_0(P), \ldots, f_m(P)$ are determined uniquely by the numbers $f_{m+1}(P), \ldots, f_{d-1}(P)$.*

Of course, all the preceding results have dual counterparts for simplicial polytopes. We only mention the dual of Theorem 17.1, the Dehn–Sommerville Relations for the simplicial d-polytopes:

Corollary 17.8. *For any simplicial d-polytope P we have*

$$\sum_{j=-1}^{d-1}(-1)^{d-1-j}\binom{j+1}{i+1}f_j(P) = f_i(P)$$

for $i = -1, \ldots, d-1$

For an entirely different approach to the Dehn–Sommerville Relations, see the remark at the end of Section 18.

§18. The Upper Bound Theorem

In this section we shall answer the following question: What is the largest number of vertices, edges, etc. of a simple d-polytope, $d \geq 3$, with a given number of facets? Moreover, we shall find out which polytopes have the largest number of vertices, edges, etc. The result which is known as the Upper Bound Theorem is a main achievement in the modern theory of convex polytopes; it was proved by McMullen in 1970.

Let us begin by noting that all simple 3-polytopes with a given number of facets have the same number of vertices and the same number of edges; this follows from the "reformulation" of the Dehn–Sommerville Relations mentioned at the beginning of Section 17:

$$f_0(P) = 2f_2(P) - 4, \qquad f_1(P) = 3f_2(P) - 6. \tag{1}$$

So, the following is only of significance for $d \geq 4$.

Recall from Section 14 that for any simplicial neighbourly d-polytope P with p vertices we have

$$f_j(P) = \binom{p}{j+1}, \qquad j = 1, \ldots, n-1.$$

(As in Section 17, we write $m := \lfloor (d-1)/2 \rfloor$ and $n := \lfloor d/2 \rfloor$.) Moreover, for these values of j, no simplicial d-polytope with p vertices can have a larger number of j-faces. So, for $1 \leq j \leq n-1$, the least upper bound for the number of j-faces of simplicial d-polytopes with p vertices equals

$$\binom{p}{j+1},$$

and this upper bound is attained by the neighbourly polytopes. Conversely, if P is a simplicial non-neighbourly d-polytope with p vertices, then

$$f_j(P) < \binom{p}{j+1}$$

for $j = n-1$, and possibly also for smaller values of j.

In the dual setting, the discussion above shows that for $m + 1 \leq j \leq d-2$, the least upper bound for the number of j-faces of a simple d-polytope with p facets equals

$$\binom{p}{d-j},$$

and that this upper bound is attained by the dual neighbourly polytopes; moreover, if P is a simple d-polytope with p facets which is not dual neighbourly, then

$$f_j(P) < \binom{p}{d-j}$$

for $j = m + 1$, and possibly also for larger values of j. The main result of this section includes these statements.

In order to state the main result, we define for $j \geq 0$

$$\Phi_j(d, p) := \sum_{i=0}^{n} \binom{i}{j}\binom{p - d + i - 1}{i} + \sum_{i=0}^{m} \binom{d - i}{j}\binom{p - d + i - 1}{i}. \quad (2)$$

With this notation the *Upper Bound Theorem* may be stated as follows:

Theorem 18.1. *For any simple d-polytope P with p facets we have*

$$f_j(P) \leq \Phi_j(d, p), \qquad j = 0, \ldots, d - 2.$$

If P is dual neighbourly, then

$$f_j(P) = \Phi_j(d, p), \qquad j = 0, \ldots, d - 2.$$

If P is not dual neighbourly, then

$$f_j(P) < \Phi_j(d, p), \qquad j = 0, \ldots, m + 1,$$

(and possibly also for larger values of j).

It is easy to verify that Theorem 18.1 holds for $d = 3$. Recall that any 3-polytope is neighbourly, and therefore any 3-polytope is also dual neighbourly. So, for $d = 3$ the statement of the theorem amounts to saying that for any simple 3-polytope P with p facets we have $f_0(P) = \Phi_0(3, p)$ and $f_1(P) = \Phi_1(3, p)$. Noting that $\Phi_0(3, p) = 2p - 4$ and $\Phi_1(3, p) = 3p - 6$, this follows immediately from (1).

Since $\Phi_{d-1}(d, p) = p$ and $\Phi_d(d, p) = 1$, the first two statements of Theorem 18.1 also hold for $j = d - 1$ and $j = d$. The proof below actually covers these values of j.

The discussion preceding Theorem 18.1 shows that we must have

$$\Phi_j(d, p) = \binom{p}{d - j}, \qquad j = m + 1, \ldots, d - 2.$$

We shall return to this matter after the proof of Theorem 18.1.

Let us also remark that Theorem 18.1 shows that the dual neighbourly polytopes are remarkably well equipped with faces: Among all simple d-polytopes with p facets, any dual neighbourly has the largest possible number of j-faces for *all* values of j between 0 and $d - 2$.

Finally, let us remark that the upper bound inequality $f_j(P) \leq \Phi_j(d, p)$ actually holds for any (i.e. not necessarily simple) d-polytope P with p facets; this is due to the fact that for any d-polytope P there is a simple d-polytope P' with the same number of facets as P and as least as many j-faces for $0 \leq j \leq d - 2$, cf. Exercise 12.4.

Proof. The proof is divided into three parts. In Part A we shall introduce certain numbers $g_i(P)$ associated with a simple d-polytope P. In Part B we shall obtain relations between the numbers $g_i(P)$ and the corresponding numbers $g_i(F)$ for facets F of P. Finally, in Part C we shall combine the results of Part A and Part B to obtain the desired conclusions.

A. In the following, let P be a simple d-polytope in $\mathbb{R}^d$, and let w be any vector in $\mathbb{R}^d$ which is admissible for P, cf. Theorem 15.1. As described in Section 15, the vector w turns the non-oriented graph $\mathscr{G}(P)$ into an oriented graph $\mathscr{G}(P, w)$. (For graph-theoretic notions, see Appendix 2.) The following is an immediate consequence of Theorem 12.12:

(a) *For each vertex x of P, the sum of the in-valence of x and the out-valence of x equals d.*

We shall need some more definitions. A *k-star*, $k = 0, \ldots, d$, is a set formed by a vertex x of P and k edges of P incident to x; the vertex x is called the *centre* of the k-star. A k-star whose edges are all oriented towards the centre is called a *k-in-star*, and a k-star whose edges are all oriented away from the centre is called a *k-out-star*.

There is a close relationship between k-faces and k-in-stars (or k-out-stars):

(b) *Let x be the centre of a k-in-star, and let F be the smallest face of P containing the k-in-star. Then F is a k-face, F is the only k-face containing the k-in-star, and x is the bottom vertex of F. The same statement with k-in-star replaced by k-out-star and bottom vertex replaced by top vertex is also valid.*

To prove (b), we first note that F is a k-face by Theorem 12.17(a). Any other face containing the k-in-star must therefore have dimension $> k$, cf. Corollary 5.5. To see that x is the bottom vertex of F, note that the only vertices of F adjacent to x are the endpoints $x_1, \ldots, x_k$ of the edges $[x, x_1], \ldots, [x, x_k]$ belonging to the k-in-star; this follows from Theorem 12.17(b). This implies that x is separated from the vertices of F adjacent to x by a suitable "horizontal" hyperplane H. Theorem 11.8, applied to F, then shows that x is the bottom vertex of F. For k-out-stars the statement is proved in a similar way.

We shall next use (b) to prove the following:

(c) *The number $f_j(P)$ of j-faces of P equals the number of j-in-stars, $j = 0, \ldots, d$.*

We shall prove (c) by showing that each j-face contains one and only one j-in-star, and each j-in-star is contained in some j-face. Let F be a j-face. Then each vertex of F is the centre of a unique j-star in F; this follows from Theorems 12.15 and 12.12. The particular j-star whose centre is the bottom vertex of F is clearly a j-in-star. On the other hand, the centre of any other j-in-star in F must also be the bottom vertex of F by (b). Hence F contains precisely one j-in-star. Finally, it follows immediately from (b) that each j-in-star in P is contained in some (in fact, a unique) j-face of P. This completes the proof of (c).

Now, for $i = 0, \ldots, d$, let $g_i(P)$ denote the number of vertices of P whose in-valence equals i. The top vertex of P has in-valence 0, and it is, in fact, the only vertex whose in-valence equals 0, cf. Theorem 15.2. Therefore:

(d) $g_0(P) = 1$.

It follows immediately from the definitions that the number of j-in-stars of P equals

$$\sum_{i=0}^{d} \binom{i}{j} g_i(P).$$

Using (c), we then obtain:

(e) $f_j(P) = \sum_{i=0}^{d} \binom{i}{j} g_i(P)$ *for* $j = 0, \ldots, d$.

Letting

$$A := \left(\binom{j}{i}\right)_{i=0,\ldots,d;\, j=0,\ldots,d}$$

we can rewrite (e) as a matrix identity,

$$\begin{pmatrix} f_0(P) \\ \vdots \\ f_d(P) \end{pmatrix} = A \begin{pmatrix} g_0(P) \\ \vdots \\ g_d(P) \end{pmatrix}.$$

Now, A is invertible. In fact, we have

$$A^{-1} = \left((-1)^{i+j}\binom{j}{i}\right)_{i=0,\ldots,d;\, j=0,\ldots,d},$$

cf. Appendix 3, (11). Therefore, the matrix identity above is equivalent to

$$\begin{pmatrix} g_0(P) \\ \vdots \\ g_d(P) \end{pmatrix} = A^{-1} \begin{pmatrix} f_0(P) \\ \vdots \\ f_d(P) \end{pmatrix},$$

whence

(f) $g_i(P) = \sum_{j=0}^{d} (-1)^{i+j} \binom{j}{i} f_j(P)$ *for* $i = 0, \ldots, d$.

In the relations (f), the right-hand sides are certainly independent of w. It then follows that, although the definition of the numbers $g_i(P)$ apparently depends on the particular choice of w, we actually have:

(g) *The numbers* $g_i(P)$, $i = 0, \ldots, d$, *are independent of* w.

It is trivial that if w is admissible for P, then $-w$ is also admissible for P. When one replaces w by $-w$, then all orientations of the edges of P are reversed. In particular, vertices having in-valence i with respect to w will have in-valence $d - i$ with respect to $-w$, cf. (a). Bearing in mind (g), it follows that

(h) $g_i(P) = g_{d-i}(P)$ *for* $i = 0, \ldots, d$.

Rewriting (e) as

$$\begin{aligned} f_j(P) &= \sum_{i=0}^{n} \binom{i}{j} g_i(P) + \sum_{i=n+1}^{d} \binom{i}{j} g_i(P) \\ &= \sum_{i=0}^{n} \binom{i}{j} g_i(P) + \sum_{i=0}^{m} \binom{d-i}{j} g_{d-i}(P), \end{aligned}$$

it then follows using (h) that

(i) $f_j(P) = \sum_{i=0}^{n} \binom{i}{j} g_i(P) + \sum_{i=0}^{m} \binom{d-i}{j} g_i(P)$ *for* $j = 0, \ldots, d$.

This relation shows that the numbers $f_j(P)$ can be expressed as non-negative linear combinations of the numbers $g_i(P)$ with i ranging only up to n. Actually, for $j \leq m + 1$ the coefficient of each $g_i(P)$ is >0 in at least one of the two sums in (i).

B. When P is a simple d-polytope, then every facet F of P is also simple, cf. Theorem 12.15. Therefore, there are numbers $g_i(F)$, $i = 0, \ldots, d - 1$, associated with F, as defined in Part A.

In the following, let F be a facet of a simple d-polytope P. Let w be admissible for P. Then, for each vertex x of F, let the *relative in-valence* of x in F be the in-valence of x in the subgraph of $\mathcal{G}(P, w)$ spanned by the vertices of F; in other words, the relative in-valence of x is the number of edges $[x, y]$ of F oriented towards x. Now, when the vector w is admissible for P, it is also admissible for F. Therefore, for any vertex x of F, the in-valence of x in $\mathcal{G}(F, w)$ equals the relative in-valence of x in F. Hence:

(j) $g_i(F)$ *equals the number of vertices of* F *whose relative in-valence is* i *for* $i = 0, \ldots, d - 1$.

Let w be admissible for P such that each vertex of P not in F is below any vertex of F, cf. Theorem 15.3. Then the relative in-valence of a vertex x of F is simply the in-valence of x in $\mathcal{G}(P, w)$. By (j), this implies

(k) $g_i(F) \leq g_i(P)$ *for* $i = 0, \ldots, d - 1$.

Suppose that for some i, we have strict inequality in (k). Then there is at least one vertex x of P not in F such that the in-valence of x is i. Therefore, the out-valence of x is $d - i$, cf. (a). It then follows that x is the centre of a unique $(d - i)$-out-star. Let G be the smallest face of P containing this $(d - i)$-out-star. Using (b), it follows that G is a $(d - i)$-face, and x is the top vertex of G. Since x is not in F, and each vertex of F is above any vertex of P not in F, we see that G and F are disjoint. Now, note that G is the intersection of the facets containing G, cf. Theorem 10.4, and the number of such facets equals i since P is simple. Let these facets be $F_1, \ldots, F_i$. Then the $i + 1$ facets $F, F_1, \ldots, F_i$ have an empty intersection since F and G are disjoint. By

Theorem 14.7, this implies that $i + 1 > n$, provided that P is a dual neighbourly polytope. In other words:

(l) *If P is a dual neighbourly polytope, then $g_i(F) = g_i(P)$ for $i = 0, \ldots, n - 1$.*

We remind the reader that in the preceding discussion, F is any facet of P. The following, therefore, is the converse of (l):

(m) *If P is not a dual neighbourly polytope, then there is a facet F of P such that $g_i(F) < g_i(P)$ for at least one $i = 0, \ldots, n - 1$.*

To prove (m), we reverse the proof of (l). If P is not a dual neighbourly polytope then there is a $k \leq n$ such that certain k facets of P, say $F_1, \ldots, F_k$, have an empty intersection, whereas any $k - 1$ facets intersect, cf. Theorem 14.7. Let G denote the intersection of $F_1, \ldots, F_{k-1}$. Then G is a face of P whose dimension equals $d - (k - 1)$, cf. Theorem 12.14. Let w be admissible for P such that any vertex of P which is not in F_k is below any vertex of F_k, cf. Theorem 15.3. Let x be the top vertex of G. Then the out-valence of x is at least $d - (k - 1)$, cf. Theorems 12.12 and 12.15, and the in-valence, therefore, is at most $k - 1$, cf. (a). Denoting the in-valence of x by i, it then follows that $i \leq n - 1$ and $g_i(F_k) < g_i(P)$.

C. Let P be a simple d-polytope with p facets, and let w be admissible for P. By an *i-incidence*, where $i = 0, \ldots, d - 1$, we shall mean a pair (F, x), where F is a facet of P and x is a vertex of F whose relative in-valence in F equals i. We denote the total number of i-incidences by I_i. It follows from (j) that

(n) $I_i = \sum_{F \in \mathscr{F}_{d-1}(P)} g_i(F)$ *for* $i = 0, \ldots, d - 1$.

Combining (n) and (k), we obtain:

(o) $I_i \leq p g_i(P)$ *for* $i = 0, \ldots, d - 1$.

Combining (n) and (l), we obtain:

(p) *If P is a dual neighbourly polytope, then $I_i = p g_i(P)$ for $i = 0, \ldots, n - 1$.*

And combining (n) and (m), we obtain:

(q) *If P is not a dual neighbourly polytope, then $I_i < p g_i(P)$ for at least one $i = 0, \ldots, n - 1$.*

We shall next prove:

(r) $I_i = (d - i) g_i(P) + (i + 1) g_{i+1}(P)$ *for* $i = 0, \ldots, d - 1$.

To obtain this, we shall determine I_i by summing over the vertices of P, rather than summing over the facets as we did in (n). Let x be a vertex of P. Then there are precisely d facets of P containing x, and by Theorem 12.12 there are also precisely d edges of P containing x. Since each facet is simple, cf.

Theorem 12.15, it follows that for each facet containing x, precisely one of the d edges containing x is not in the facet, cf. Theorem 12.12; we shall call this edge the *external edge* of the facet. Note that, conversely, each edge containing x is the external edge of some facet containing x; this follows immediately from Theorem 12.17. Now, for a facet F containing the vertex x, the pair (F, x) is an i-incidence if and only if one of the following two conditions hold:

(α) *x has in-valence i in P, and the external edge of F is oriented away from x.*
(β) *x has in-valence $i + 1$ in P, and the external edge of F is oriented towards x.*

If x has in-valence i, then there are $d - i$ facets F such that (α) holds. If x has in-valence $i + 1$, then there are $i + 1$ facets F such that (β) holds. This proves (r).

We next combine (o) and (r) to obtain:

$$g_{i+1}(P) \leq \frac{p - d + i}{i + 1} g_i(P), \qquad i = 0, \ldots, d - 1.$$

Since $g_0(P) = 1$, cf. (d), it follows by induction that

(s) $g_i(P) \leq \binom{p - d + i - 1}{i}$ *for* $i = 0, \ldots, d$.

In a similar way, combining (p) and (r), we obtain:

(t) *If P is a dual neighbourly polytope, then*

$$g_i(P) = \binom{p - d + i - 1}{i}$$

for $i = 0, \ldots, n$.

And, combining (q) and (r), we get:

(u) *If P is not a dual neighbourly polytope, then*

$$g_i(P) < \binom{p - d + i - 1}{i}$$

for at least one $i = 0, \ldots, n$.

We can now complete the proof. Combining (i) and (s) we obtain

$$f_j(P) \leq \Phi_j(d, p), \qquad j = 0, \ldots, d,$$

proving the first statement of the theorem. Combining (i) and (t) we get

$$f_j(P) = \Phi_j(d, p), \qquad j = 0, \ldots, d,$$

when P is a dual neighbourly polytope, proving the second statement of the theorem. Finally, as remarked earlier, for $0 \leq j \leq m + 1$ the coefficient of each $g_i(P)$ is >0 in at least one of the two sums in (i). Therefore, combining (i) and (u) we obtain the third statement of the theorem. □

The proof above of the Upper Bound Theorem only contains one computation, namely, the use of formula (11) from Appendix 3 leading to (f). However, in the proof we do not really need (f), we only need to know that $g_i(P)$ can be expressed by the numbers $f_j(P)$ in some way. To see this it suffices to know that the matrix A is invertible, and that follows immediately from the fact that A is an upper triangular matrix with 1's in the diagonal. The explicit formula (f) is only included here because of its relevance to the discussion in Section 20.

As mentioned just after the statement of Theorem 18.1, the expression for $\Phi_j(d, p)$ can be simplified for $m + 1 \leq j \leq d - 2$. Moreover, for $j = 0$ we have a very simple expression for $\Phi_j(d, p)$, and for the remaining values of j we have a reformulation which may occasionally be useful:

Theorem 18.2. (a) *The value of* $\Phi_0(d, p)$ *equals*

$$\binom{p - m - 1}{n} + \binom{p - n - 1}{m}.$$

(b) *For* $j = 1, \ldots, n$, *the value of* $\Phi_j(d, p)$ *equals*

$$\binom{p}{d - j} + \binom{p - d + j - 1}{j}\binom{p - d + n}{n - j} - \sum_{i=m+1}^{d-j}\binom{d - i}{j}\binom{p - d + i - 1}{i}.$$

(c) *For* $j = m + 1, \ldots, d - 2$, *the value of* $\Phi_j(d, p)$ *equals*

$$\binom{p}{d - j}.$$

Proof. (a) By the definition (2) we have

$$\Phi_0(d, p) = \sum_{i=0}^{n}\binom{p - d + i - 1}{i} + \sum_{i=0}^{m}\binom{p - d + i - 1}{i}.$$

The desired expression then follows easily using Appendix 2, (9).

(b) We can rewrite the first sum in (2) using identities from Appendix 3 as indicated:

$$\sum_{i=0}^{n}\binom{i}{j}\binom{p - d + i - 1}{i}$$

$$\overset{(2)}{=} \sum_{i=0}^{n}(-1)^i\binom{i}{j}\binom{d - p}{i} \overset{(10)}{=} (-1)^n\binom{d - p}{j}\binom{d - p - j - 1}{n - j}$$

$$\overset{(2)}{=} (-1)^n(-1)^j\binom{-d + p + j - 1}{j}(-1)^{n-j}\binom{-d + p + j + 1 + n - j - 1}{n - j}$$

$$= \binom{p - d + j - 1}{j}\binom{p - d + n}{n - j}.$$

Hence, we have the second term in the desired expression.

We next prove that for $0 \leq j \leq d$ we have

$$\sum_{i=0}^{d-j} \binom{d-i}{j}\binom{p-d+i-1}{i} = \binom{p}{d-j}. \tag{3}$$

The validity of (3) is proved using identities from Appendix 3 as indicated:

$$\begin{aligned}
\sum_{i=0}^{d-j} \binom{d-i}{j}\binom{p-d+i-1}{i} &\overset{(3)}{=} \sum_{i=0}^{d-j} (-1)^{d-i-j}\binom{-j-1}{d-i-j}\binom{p-d+i-1}{i} \\
&\overset{(2)}{=} \sum_{i=0}^{d-j} (-1)^{d-i-j}\binom{-j-1}{d-i-j}(-1)^i\binom{-p+d}{i} \\
&= (-1)^{d-j}\sum_{i=0}^{d-j} \binom{-p+d}{i}\binom{-j-1}{(d-j)-i} \\
&\overset{(7)}{=} (-1)^{d-j}\binom{-p+d-j-1}{d-j} \\
&\overset{(2)}{=} \binom{p}{d-j},
\end{aligned}$$

as desired.

Using (3) we can now rewrite the second sum in (2):

$$\begin{aligned}
&\sum_{i=0}^{m}\binom{d-i}{j}\binom{p-d+i-1}{i} \\
&\quad = \sum_{i=0}^{d-j} \binom{d-i}{j}\binom{p-d+i-1}{i} - \sum_{i=m+1}^{d-j} \binom{d-i}{j}\binom{p-d+i-1}{i} \\
&\quad = \binom{p}{d-j} - \sum_{i=m+1}^{d-j} \binom{d-i}{j}\binom{p-d+i-1}{i}.
\end{aligned}$$

Hence, we have the two remaining terms in the desired expression.

(c) Although we already know that the statement is true, we would like to give a direct proof. For $j \geq n + 1$, each term in the first sum in (2) has the value 0. In the second sum, all terms corresponding to values of i that are $> d - j$ also have the value 0. Therefore,

$$\Phi_j(d, p) = \sum_{i=0}^{d-j} \binom{d-i}{j}\binom{p-d+i-1}{i}.$$

Combining with (3) above, we then get

$$\Phi_j(d, p) = \binom{p}{d-j}.$$

When $m = n$, this completes the proof. When $m = n - 1$, it remains to consider the value $j = m + 1 = n$. However, this is easily handled by returning to the expression for $\Phi_j(d, p)$ in case (b). The details are left to the reader. □

By duality, we also have an Upper Bound Theorem for the simplicial d-polytopes. It may be stated as follows:

Corollary 18.3. *For any simplicial d-polytope P with p vertices we have*

$$f_j(P) \leq \Phi_{d-1-j}(d, p), \qquad j = 1, \ldots, d-1.$$

If P is neighbourly, then

$$f_j(P) = \Phi_{d-1-j}(d, p), \qquad j = 1, \ldots, d-1.$$

If P is not neighbourly, then

$$f_j(P) < \Phi_{d-1-j}(d, p), \qquad j = n-1, \ldots, d-1,$$

(and possibly also for smaller values of j).

Finally, it is interesting to note that (f) and (h) in the proof of Theorem 18.1 show that

$$\sum_{j=0}^{d} (-1)^j \binom{j}{i} f_j(P) = \sum_{j=0}^{d} (-1)^{d+j} \binom{j}{d-i} f_j(P), \qquad i = 0, \ldots, d,$$

i.e. $(f_0(P), \ldots, f_{d-1}(P))$ satisfies the Dehn–Sommerville System of Theorem 17.5. Hence, we have an independent proof of the Dehn–Sommerville Relations which does not rely on Euler's Relation.

§19. The Lower Bound Theorem

In the preceding section we determined the largest number of vertices, edges, etc. of a simple d-polytope, $d \geq 3$, with a given number of facets. In this section we shall find the smallest number of vertices, edges, etc. The result which is known as the Lower Bound Theorem was proved by Barnette in 1971–73. Like the Upper Bound Theorem, it is a main achievement in the modern theory of convex polytopes.

As we saw at the beginning of Section 18, all simple 3-polytopes with a given number of facets have the same number of vertices and the same number of edges. So, as in the case of the Upper Bound Theorem, the problem is only of significance for $d \geq 4$.

We define

$$\varphi_j(d, p) := \begin{cases} (d-1)p - (d+1)(d-2), & j = 0; \\ \binom{d}{j+1} p - \binom{d+1}{j+1}(d-1-j), & j = 1, \ldots, d-2. \end{cases}$$

Note that

$$\varphi_{d-2}(d, p) = dp - \binom{d+1}{d-1}$$
$$= dp - (d^2 + d)/2.$$

With this notation the *Lower Bound Theorem* may be stated as follows:

Theorem 19.1. *For any simple d-polytope P with p facets we have*

$$f_j(P) \geq \varphi_j(d, p), \qquad j = 0, \ldots, d-2.$$

Moreover, there are simple d-polytopes P with p facets such that

$$f_j(P) = \varphi_j(d, p), \qquad j = 0, \ldots, d-2.$$

Since $\varphi_0(3, p) = 2p - 4$ and $\varphi_1(3, p) = 3p - 6$, we see immediately as in the case of the Upper Bound Theorem that the theorem is true for $d = 3$, in fact, with equality for all simple polytopes.

Before proving Theorem 19.1 we need some notation and some preparatory lemmas.

We remind the reader that a facet system in a polytope P is a non-empty set $\mathscr{S}$ of facets of P. When $\mathscr{S}$ is a facet system in P, we denote by $\mathscr{G}(\mathscr{S})$ the union of the subgraphs $\mathscr{G}(F)$, $F \in \mathscr{S}$, of $\mathscr{G}(P)$, and we say that $\mathscr{S}$ is connected if $\mathscr{G}(\mathscr{S})$ is a connected graph. These concepts were introduced in Section 15, where we also proved some important results about connectedness properties of $\mathscr{G}(\mathscr{S})$.

When $\mathscr{S}$ is a facet system in P and G is a face of P, then we shall say that G is in $\mathscr{S}$ or G is a face of $\mathscr{S}$, if G is a face of some facet F belonging to $\mathscr{S}$. In particular, the vertices of $\mathscr{S}$ are the vertices of the facets in $\mathscr{S}$.

In the following, we shall restrict our attention to facet systems in simple polytopes. Let $\mathscr{S}$ be a facet system in a simple d-polytope P, and let x be a vertex of $\mathscr{S}$. Then x is a vertex of at least one member F of $\mathscr{S}$. Therefore, the $d - 1$ edges of F incident to x are edges of $\mathscr{S}$. If the remaining edge of P incident to x is also in $\mathscr{S}$, we shall say that x is *internal* in $\mathscr{S}$ or that x is an internal vertex of $\mathscr{S}$. If, on the other hand, the remaining edge of P incident to x is not in $\mathscr{S}$, we shall say that x is *external* in $\mathscr{S}$ or that x is an external vertex of $\mathscr{S}$. In other words, a vertex x of $\mathscr{S}$ is external if and only if it is a vertex of only one member of $\mathscr{S}$.

The first lemma ensures the existence of external vertices under an obvious condition. (In the following, we actually need only the existence of just one external vertex.)

Lemma 19.2. *Let $\mathscr{S}$ be a facet system in a simple d-polytope P such that at least one vertex of P is not in $\mathscr{S}$. Then $\mathscr{S}$ has at least d external vertices.*

PROOF. If all vertices of $\mathscr{S}$ are external, then each member of $\mathscr{S}$ contributes at least d external vertices. Suppose that some vertex z of $\mathscr{S}$ is internal. By the assumption we also have a vertex y not in $\mathscr{S}$. We then use the d-connectedness of $\mathscr{G}(P)$, cf. Theorem 15.6, to get d independent paths joining y and z. Traversing the ith path from y to z, let x_i be the first vertex which is in $\mathscr{S}$. Then the preceding edge is not in $\mathscr{S}$, and therefore x_i is external in $\mathscr{S}$. Since the x_i's are distinct, we have the desired conclusion. □

During the proof of Theorem 15.7 it was shown that if $\mathscr{S}$ is a connected facet system in P and $\mathscr{S}$ has at least two members, then there is a member F_0 of $\mathscr{S}$ such that $\mathscr{S}\backslash\{F_0\}$ is again connected. When P is simple, we have the following much stronger result:

Lemma 19.3. *Let $\mathscr{S}$ be a connected facet system in a simple d-polytope P. Assume that at least one vertex of P is not in $\mathscr{S}$, and that $\mathscr{S}$ has at least two members. Then there is a pair (x_0, F_0) formed by an external vertex x_0 of $\mathscr{S}$ and the unique member F_0 of $\mathscr{S}$ containing x_0 such that the facet system $\mathscr{S}\backslash\{F_0\}$ is again connected.*

PROOF. We know from Lemma 19.2 that $\mathscr{S}$ has external vertices. Let (x_1, F_1) be a pair formed by an external vertex x_1 of $\mathscr{S}$ and the unique member F_1 of $\mathscr{S}$ containing x_1. Suppose that $\mathscr{S}\backslash\{F_1\}$ is not connected. Let $\mathscr{S}_1$ be a maximal connected subsystem of $\mathscr{S}\backslash\{F_1\}$. We shall prove that then there is another pair (x_2, F_2) such that $\mathscr{S}_1 \cup \{F_1\}$ is a connected subsystem of $\mathscr{S}\backslash\{F_2\}$. In other words: if $\mathscr{S}\backslash\{F_1\}$ is not connected, then we can replace (x_1, F_1) by some (x_2, F_2) in such a manner that the maximum number of members of a connected subsystem of $\mathscr{S}\backslash\{F_2\}$ is larger than the maximum number of members of a connected subsystem of $\mathscr{S}\backslash\{F_1\}$. Continuing this procedure eventually leads to a pair (x_0, F_0) with the property that $\mathscr{S}\backslash\{F_0\}$ is connected.

Now, let (x_1, F_1) and $\mathscr{S}_1$ be as explained above. We first prove that $\mathscr{S}_1 \cup \{F_1\}$ is connected. Let y be any vertex of $\mathscr{S}_1$; note that $y \neq x_1$ since F_1 is the only member of $\mathscr{S}$ containing x_1 and $F_1 \notin \mathscr{S}_1$. By the connectedness of $\mathscr{S}$ there is a path in $\mathscr{G}(\mathscr{S})$ joining y and x_1. Traversing this path from y to x_1, let F be a member of $\mathscr{S}$ containing the first edge of the path not in $\mathscr{S}_1$. (Since x_1 is not in $\mathscr{S}_1$, such an edge certainly exists.) Then clearly $\mathscr{S}_1 \cup \{F\}$ is connected. By the maximality property of $\mathscr{S}_1$ we must have $F = F_1$, whence $\mathscr{S}_1 \cup \{F_1\}$ is connected, as desired. Let $\mathscr{S}'_1 := \mathscr{S}\backslash(\mathscr{S}_1 \cup \{F_1\})$. Then $\mathscr{S}'_1$ is non-empty, possibly disconnected. By Lemma 19.2, $\mathscr{S}'_1$ has external vertices. Not every external vertex of $\mathscr{S}'_1$ can be in F_1. For then every path joining a vertex of $\mathscr{S}'_1$ and a vertex of P not in $\mathscr{S}'_1$ would have to pass through a vertex of F_1, whence the subgraph of $\mathscr{G}(P)$ spanned by ext $P\backslash$ext F_1 would be disconnected, contradicting Theorem 15.5. Let x_2 be an external vertex of $\mathscr{S}'_1$ not in F_1, and let F_2 be the unique member of $\mathscr{S}'_1$ containing x_2. Then actually x_2 is external in $\mathscr{S}$. For if not, then x_2 would have to be a vertex of some member F of $\mathcal{S}_1$, since F_2 is the only member of $\mathcal{S}'_1$ containing x_2, and x_2 is not in F_1; but then $\mathcal{S}_1 \cup \{F_2\}$ would be connected, contradicting the maximality property of $\mathcal{S}_1$. Hence, x_2 is external in $\mathcal{S}$, the facet F_2 is the unique member of $\mathcal{S}$ containing x_2, and $\mathcal{S}_1 \cup \{F_1\}$ is a connected subsystem of $\mathcal{S}\setminus\{F_2\}$, as desired. □

Lemma 19.4. *Let $\mathcal{S}$ be a connected facet system in a simple d-polytope P. Assume that at least one vertex of P is not in $\mathcal{S}$, and that $\mathcal{S}$ has at least two members. Let (x_0, F_0) be as in Lemma 19.3. Then at least $d - 1$ vertices of P are internal in $\mathcal{S}$ but external in $\mathcal{S}\setminus\{F_0\}$.*

PROOF. By the connectedness of $\mathcal{S}$, there is a member F of $\mathcal{S}$ with $F \neq F_0$ and $F \cap F_0 \neq \emptyset$. Then by Theorem 12.14, the face $F \cap F_0$ has dimension $d - 2$, whence F and F_0 have at least $d - 1$ vertices in common. Being vertices of two members of $\mathcal{S}$, such $d - 1$ vertices are all internal in $\mathcal{S}$. So, if they are all external in $\mathcal{S}\setminus\{F_0\}$, we have the desired conclusion. If they are not all external in $\mathcal{S}\setminus\{F_0\}$, one of the vertices, say y, is internal in $\mathcal{S}\setminus\{F_0\}$. In particular, $y \neq x_0$. Then by Theorem 15.7 there are $d - 1$ independent paths in $\mathcal{G}(\mathcal{S})$ joining x_0 and y. Traversing the ith path from x_0 to y, let x_i be the first vertex which is in $\mathcal{S}\setminus\{F_0\}$. Then the preceding edge $[u_i, x_i]$ is not in $\mathcal{S}\setminus\{F_0\}$, and therefore x_i is external in $\mathcal{S}\setminus\{F_0\}$. In particular, $x_i \neq x_0$ and $x_i \neq y$. Moreover, since $[u_i, x_i]$ is not in $\mathcal{S}\setminus\{F_0\}$, it must be in F_0, whence x_i is a vertex of F_0. Since x_i is also a vertex of $\mathcal{S}\setminus\{F_0\}$, we see that x_i belongs to at least two members of $\mathcal{S}$, showing that x_i is internal in $\mathcal{S}$. In conclusion, the $d - 1$ vertices $x_1, \ldots, x_{d-1}$ are internal in $\mathcal{S}$ but external in $\mathcal{S}\setminus\{F_0\}$. □

Lemma 19.5. *Let $\mathcal{S}$ be a facet system in a simple d-polytope P such that at least one vertex of P is not in $\mathcal{S}$. Then there are at least d facets $G_1, \ldots, G_d$ of P such that $G_1, \ldots, G_d$ are not in $\mathcal{S}$ but each contains a $(d - 2)$-face which is in $\mathcal{S}$.*

PROOF. Let x be a vertex of P not in $\mathcal{S}$. Let Q be a dual of P in $\mathbb{R}^d$, and let ψ be an anti-isomorphism from $(\mathcal{F}(P), \subset)$ onto $(\mathcal{F}(Q), \subset)$. Writing

$$\mathcal{S} = \{F_1, \ldots, F_m\},$$

x is not a vertex of any of the F_i's, whence the facet $\psi(\{x\})$ of Q does not contain any of the vertices $\psi(F_i)$ of Q, cf. Theorem 9.8. Let z be a point of $\mathbb{R}^d$ outside Q but "close" to $\psi(\{x\})$ such that every vertex of Q is also a vertex of $Q' := \operatorname{conv}(Q \cup \{z\})$; then the vertices of Q' are the vertices of Q plus the vertex z and the edges of Q' are the edges of Q plus the edges $[z, u]$, where $u \in \operatorname{ext} \psi(\{x\})$. (Supposing that $o \in \operatorname{int} P$, one may take Q' to be the polar of a polytope obtained by truncating the vertex x of P, cf. Section 11.) By Theorem 15.6 there are d independent paths in $\mathcal{G}(Q')$ joining the vertices z and $\psi(F_1)$. Traversing the ith path from z to $\psi(F_1)$, let y_i be the vertex preceding the first of any of the vertices $\psi(F_1), \ldots, \psi(F_m)$ on the path. Then by duality,

$\psi^{-1}(\{y_1\}), \ldots, \psi^{-1}(\{y_d\})$ are d facets of P not in $\mathscr{S}$, each having a $(d-2)$-face in common with some member of $\mathscr{S}$. □

We are now in position to prove the Lower Bound Theorem:

Proof (Theorem 19.1). We divide the proof into four parts. In Part A we prove the inequality for $j = 0$, and in Part B we prove the inequality for $j = d - 2$; here the lemmas above are used. In Part C we cover the remaining values of j; the proof is by induction. Finally, in Part D we exhibit polytopes for which we have equality.

A. We choose a vertex x of P and let

$$\mathscr{S} := \{F \in \mathscr{F}_{d-1}(P) \mid x \notin F\}.$$

Then $\mathscr{G}(\mathscr{S})$ is the subgraph of $\mathscr{G}(P)$ spanned by ext $P \setminus \{x\}$, whence, by Theorem 15.5, $\mathscr{S}$ is a connected facet system. The number of members of $\mathscr{S}$ is $p - d$.

Only one vertex of P is not in $\mathscr{S}$, namely, the vertex x. The d vertices of P adjacent to x are external vertices of $\mathscr{S}$, and they are the only external vertices of $\mathscr{S}$. Hence, the number of internal vertices of $\mathscr{S}$ is $f_0(P) - (d + 1)$.

If $p = d + 1$, then P is a d-simplex and the inequality holds with equality. If $p \geq d + 2$, we remove facets from $\mathscr{S}$ one by one by successive applications of Lemma 19.3. At each removal, at least $d - 1$ vertices change their status from internal to external by Lemma 19.4. After $p - d - 1$ removals, we end up with a one-membered facet system. The total number of vertices which during the removal process have changed their status is therefore at least

$$(p - d - 1)(d - 1).$$

Since the number of internal vertices equals $f_0(P) - (d + 1)$, it follows that

$$f_0(P) - (d + 1) \geq (p - d - 1)(d - 1),$$

whence

$$f_0(P) \geq (d - 1)p - (d + 1)(d - 2),$$

as desired.

B. This part is divided into two steps. We first prove that if there is a constant K depending on d only such that

$$f_{d-2}(P) \geq d f_{d-1}(P) - K \tag{1}$$

for all simple d-polytopes P, then the desired inequality

$$f_{d-2}(P) \geq d f_{d-1}(P) - (d^2 + d)/2 \tag{2}$$

must hold. Then, in the second step, we show that (1) holds with $K = d^2 + d$.

Suppose that the inequality (2) does not hold in general. Then there is a simple d-polytope P in $\mathbb{R}^d$ such that

$$f_{d-2}(P) = df_{d-1}(P) - (d^2 + d)/2 - r$$

for some $r > 0$. Let Q be a dual of P in $\mathbb{R}^d$. Then Q is a simplicial d-polytope with

$$f_1(Q) = df_0(Q) - (d^2 + d)/2 - r.$$

By Theorem 11.10 we may assume that there is a facet F of Q such that the orthogonal projection of $\mathbb{R}^d$ onto the hyperplane aff F maps $Q \backslash F$ into ri F. Let Q' denote the polytope obtained by reflecting Q in aff F. Then $Q_1 := Q \cup Q'$ is again a d-polytope by the property of F. It is clear that Q_1 is simplicial. Since F has d vertices, we have

$$f_0(Q_1) = 2f_0(Q) - d,$$

and since F has

$$\binom{d}{2} = (d^2 - d)/2$$

edges, we have

$$f_1(Q_1) = 2f_1(Q) - (d^2 - d)/2.$$

We then get

$$\begin{aligned} f_1(Q_1) &= 2(df_0(Q) - (d^2 + d)/2 - r) - (d^2 - d)/2 \\ &= df_0(Q_1) - (d^2 + d)/2 - 2r. \end{aligned}$$

Let P_1 be a dual of Q_1. Then P_1 is a simple d-polytope with

$$f_{d-2}(P_1) = df_{d-1}(P_1) - (d^2 + d)/2 - 2r$$

This shows that P_1 fails to satisfy (2) by at least $2r$ faces of dimension $d - 2$. Continuing this construction we conclude that no inequality of the form (1) can hold for all simple d-polytopes. This completes the first step.

To carry out the second step, let P be any simple d-polytope, and let $p := f_{d-1}(P)$. Let x and $\mathscr{S}$ be as in Part A. If $p = d + 1$, then P is a d-simplex, whence

$$\begin{aligned} f_{d-2}(P) &= \binom{d+1}{d-1} \\ &= d(d + 1) - (d^2 + d)/2 \\ &= dp - (d^2 + d)/2 \\ &> dp - (d^2 + d), \end{aligned}$$

as desired. For $p \geq d + 2$, we shall remove the facets in $\mathscr{S}$ one by one by successive applications of Lemma 19.3 as we did in Part A. Let F_i denote the

ith member of $\mathscr{S}$ to be removed, let x_i denote a corresponding external vertex of $\mathscr{S}\setminus\{F_1, \ldots, F_{i-1}\}$ contained in F_i, and let

$$\mathscr{S}_i := \{F_i \cap F_j | F_i \cap F_j \neq \emptyset, j = i + 1, \ldots, p\}, \qquad i = 1, \ldots, p - d - 1.$$

Then $\mathscr{S}_i$ is a facet system in F_i, cf. Theorem 12.14.

Now, let us say that a $(d - 2)$-face G of F_i is of type 1 in F_i if G is not in $\mathscr{S}_i$ but some $(d - 3)$-face of G is in $\mathscr{S}_i$. Lemma 19.5 can be applied to the facet system $\mathscr{S}_i$ in F_i, for x_i is a vertex of F_i not in $\mathscr{S}_i$. As a result we get $d - 1$ $(d - 2)$-faces of F_i of type 1. Note that a $(d - 2)$-face of type 1 in F_i is not a face of any F_j with $j > i$.

For $i = 1, \ldots, p - d - 1$, let

$$q_i := \max\{j | i < j, F_i \cap F_j \neq \emptyset\}.$$

Then $G_i := F_i \cap F_{q_i}$ is a $(d - 2)$-face of F_i which we shall call a $(d - 2)$-face of type 2 in F_i. Note that F_i and F_{q_i} are the only facets of P containing G_i, cf. Theorem 12.14, that G_i is not at the same time of type 1 in F_i, and that G_i is neither of type 1 nor type 2 in F_{q_i}.

The discussion above now shows that for $i = 1, \ldots, p - d - 1$, the number of $(d - 2)$-faces contributed by F_i is at least d, namely, $d - 1$ of type 1 and one of type 2. Therefore, the total number of $(d - 2)$-faces of P is at least

$$(p - d - 1)d = dp - (d^2 + d),$$

as desired.

C. Using induction on d we shall prove that the inequality holds for the remaining values of j, namely, $j = 1, \ldots, d - 3$. We first note that for $d = 3$ there are no such remaining values; this ensures the start of the induction. So, let $d \geq 4$ and assume that the inequality holds for dimension $d - 1$ and $j = 1, \ldots, (d - 1) - 3$. Let P be a simple d-polytope with p facets, and let j have any of the values $1, \ldots, d - 3$. By a *j-incidence* we shall mean a pair (F, G) where F is a facet of P and G is a j-face of F. (This notion of incidence differs from the one used in the proof of the Upper Bound Theorem.) It is clear that the number of j-incidences equals

$$\sum_{F \in \mathscr{F}_{d-1}(P)} f_j(F).$$

Moreover, since each j-face of P is contained in precisely $d - j$ facets, the number of j-incidences also equals $(d - j)f_j(P)$. Hence,

$$(d - j)f_j(P) = \sum_{F \in \mathscr{F}_{d-1}(P)} f_j(F). \tag{3}$$

We next note that for any facet F of P we have

$$f_j(F) \geq \binom{d-1}{j+1} f_{d-2}(F) - \binom{d}{j+1}(d - 2 - j); \tag{4}$$

in fact, for $j = 1, \ldots, d - 4$ this follows from the induction hypothesis applied to F, and for $j = d - 3$ it follows from the result of Part B applied to F. Combining (3) and (4) we obtain

$$\begin{aligned}(d - j)f_j(P) &\geq \sum_{F \in \mathscr{F}_{d-1}(P)} \left(\binom{d-1}{j+1} f_{d-2}(F) - \binom{d}{j+1}(d - 2 - j) \right) \\ &= \binom{d-1}{j+1} \sum_{F \in \mathscr{F}_{d-1}(P)} f_{d-2}(F) - \binom{d}{j+1}(d - 2 - j) \sum_{F \in \mathscr{F}_{d-1}(P)} 1 \\ &= \binom{d-1}{j+1} \sum_{F \in \mathscr{F}_{d-1}(P)} f_{d-2}(F) - \binom{d}{j+1}(d - 2 - j)p.\end{aligned}$$

Here

$$\sum_{F \in \mathscr{F}_{d-1}(P)} f_{d-2}(F) = 2f_{d-2}(P)$$

since each $(d - 2)$-face of P is contained in precisely two facets. Hence,

$$(d - j)f_j(P) \geq \binom{d-1}{j+1} 2f_{d-2}(P) - \binom{d}{j+1}(d - 2 - j)p.$$

We next apply the result of Part B to P, obtaining

$$(d - j)f_j(P) \geq \binom{d-1}{j+1} 2\left(dp - \binom{d+1}{d-1}\right) - \binom{d}{j+1}(d - 2 - j)p.$$

An easy calculation shows that the right-hand side of this inequality may be rewritten as

$$(d - j)\left(\binom{d}{j+1}p - \binom{d+1}{j+1}(d - 1 - j)\right).$$

Cancelling the factor $d - j$, we obtain the desired inequality.

D. It is easy to see that we have equality for all j when P is a d-simplex. Truncation of one vertex of a simple d-polytope P with p facets produces a simple d-polytope P' with $p + 1$ facets, with

$$\binom{d}{j+1}$$

more j-faces than P for $1 \leq j \leq d - 2$, and with $d - 1$ more vertices than P, cf. Theorem 12.18. It is easy to see that if we have equality for P, then we also have equality for P'. Hence, the desired polytopes may be obtained from a d-simplex by repeated truncation of vertices. This completes the proof of Theorem 19.1. □

It would be desirable to have a more direct proof of the Lower Bound Inequalities than the one given in Parts A, B and C above. As a beginning, one could think of a direct proof of the inequality for $j = d - 2$, replacing the two-step proof of Part B. In the second step we proved that (1) holds with $K = d^2 + d$. Compared to the desired inequality, the deficit amounts to $(d^2 + d)/2$. However, when counting the $(d - 2)$-faces we did not count those containing x; the number of such $(d - 2)$-faces equals

$$\binom{d}{d-2} = (d^2 - d)/2.$$

This improvement does not yield the desired inequality, but it reduces the deficit to d.

In Part D of the proof of Theorem 19.1, we showed that we have equality for the *truncation polytopes*, i.e. the polytopes obtained from simplices by successive truncations of vertices. For $d \geq 4$ it is known that if $f_j(P) = \varphi_j(d, p)$ for just one value of j, then P must be a truncation polytope. For $d = 3$ the situation is different. As we know, all simple 3-polytopes yield equality. On the other hand, there are simple 3-polytopes which are not truncation polytopes, for example, the parallellotopes.

In Section 18 it was indicated that the upper bound $\Phi_j(d, p)$ is also valid for non-simple polytopes. In contrast to this, little seems to be known about lower bounds for non-simple polytopes.

In its dual form, the Lower Bound Theorem may be stated as follows:

Corollary 19.6. *For any simplicial d-polytope P with p vertices we have*

$$f_j(P) \geq \varphi_{d-1-j}(d, p), \qquad j = 1, \ldots, d - 1.$$

Moreover, there are simplicial d-polytopes P with p vertices such that

$$f_j(P) = \varphi_{d-1-j}(d, p), \qquad j = 1, \ldots, d - 1.$$

Equality in Corollary 19.7 is attained by the duals of the truncation polytopes, and, for $d \geq 4$, only by these. They are the polytopes obtained from simplices by successive addition of pyramids over facets; they are called *stacked polytopes*.

It is interesting to note that the Lower Bound Inequalities are closely related to inequalities between the numbers $g_i(P)$ introduced in Section 18. For details, see Section 20.

§20. McMullen's Conditions

At the beginning of Section 16 it was indicated that it is not known how to characterize the f-vectors of d-polytopes among all d-tuples of positive integers. However, the more restricted problem of characterizing the f-vectors of simple (or simplicial) d-polytopes has recently been solved. It was conjectured by McMullen in 1971 that a certain set of three conditions would characterize the f-vectors of simplicial d-polytopes. In 1980, the sufficiency of McMullen's conditions was established by Billera and Lee, and the necessity was established by Stanley. We shall briefly report on these fundamental results, but we shall not be able to include the proofs. As in the preceding sections, we shall express ourselves in terms of simple (rather than simplicial) polytopes.

We begin by introducing some notation. For any d-tuple $f = (f_0, \ldots, f_{d-1})$ of positive integers we define

$$g_i(f) := \sum_{j=0}^{d} (-1)^{i+j} \binom{j}{i} f_j, \qquad i = 0, \ldots, d,$$

where, by convention, we always put $f_d = 1$. Note that when f is the f-vector of a simple d-polytope P, then $g_i(f)$ is just the well-known number $g_i(P)$, cf. statement (f) in the proof of Theorem 18.1. Note also that by the argument leading to this statement,

$$f_j = \sum_{i=0}^{d} \binom{i}{j} g_i(f), \qquad j = 0, \ldots, d. \tag{1}$$

We need some more notation. Let h and k be positive integers. Then using induction on k, it is easy to see that there exist uniquely determined positive integers $r_0, r_1, \ldots, r_q$ such that

$$r_0 > r_1 > \cdots > r_q \geq k - q \geq 1$$

and

$$h = \binom{r_0}{k} + \binom{r_1}{k-1} + \cdots + \binom{r_q}{k-q}. \tag{2}$$

In fact, r_0 is the largest integer such that

$$h \geq \binom{r_0}{k}, \tag{3}$$

r_1 is the largest integer such that

$$h - \binom{r_0}{k} \geq \binom{r_1}{k-1}$$

(unless we have equality in (3) in which case $q = 0$), etc. The representation (2) is called the *k-canonical representation* of h.

Given the k-canonical representation (2) of h, we define the kth *pseudo-power* $h^{\langle k \rangle}$ of h by

$$h^{\langle k \rangle} := \binom{r_0 + 1}{k + 1} + \binom{r_1 + 1}{k} + \cdots + \binom{r_q + 1}{k - q + 1}. \tag{4}$$

Note that at the same time, (4) is the $(k + 1)$-canonical representation of $h^{\langle k\rangle}$. It is easy to see that the kth pseudopower is *monotone*:

$$h_1 < h_2 \Rightarrow h_1^{\langle k\rangle} < h_2^{\langle k\rangle}.$$

The definition of $h^{\langle k\rangle}$ is extended to $h = 0$ by letting

$$0^{\langle k\rangle} := 0.$$

We can now formulate the characterization; recall that $m := \lfloor (d-1)/2 \rfloor$ and $n := \lfloor d/2 \rfloor$. The conditions (a)–(c) are *McMullen's Conditions.*

Theorem 20.1. *A d-tuple $f = (f_0, \ldots, f_{d-1})$ of positive integers is the f-vector of a simple d-polytope if and only if the following three conditions hold:*

(a) $g_i(f) = g_{d-i}(f)$ *for* $i = 0, \ldots, m$.

(b) $g_i(f) \le g_{i+1}(f)$ *for* $i = 0, \ldots, n-1$.

(c) $g_{i+1}(f) - g_i(f) \le (g_i(f) - g_{i-1}(f))^{\langle i\rangle}$ *for* $i = 1, \ldots, n-1$.

We know from the preceding sections that the Dehn–Sommerville Relations, the Upper Bound Inequalities and the Lower Bound Inequalities hold for f-vectors of simple polytopes. Therefore, it follows from Theorem 20.1 that if (a)–(c) hold for some $f = (f_0, \ldots, f_{d-1})$, then the Dehn–Sommerville Relations, the Upper Bound Inequalities and the Lower Bound Inequalities must also hold for f. We shall see how this can be demonstrated. (This may give the reader some idea of the significance of the conditions (a)–(c).)

We first note that by the definition of $g_i(f)$, condition (a) is equivalent to saying that $(f_0, \ldots, f_{d-1})$ satisfies the Dehn–Sommerville System of Theorem 17.5 which in turn is equivalent to saying that $(f_0, \ldots, f_{d-1})$ satisfies the Dehn–Sommerville Relations.

To deduce the Upper Bound and Lower Bound Inequalities we need the observations that

$$g_d(f) = 1$$

and

$$g_{d-1}(f) = f_{d-1} - d.$$

By (a), we then also have

$$g_0(f) = 1$$

and

$$g_1(f) = f_{d-1} - d.$$

We begin with the Lower Bound Inequalities. By (a) and (b),

$$g_1(f) \le g_i(f), \qquad i = 1, \ldots, d-1. \tag{5}$$

Using (1), (5) and the observations above, we have

$$\begin{aligned} f_0 &= \sum_{i=0}^{d} \binom{i}{0} g_i(f) \\ &= g_0(f) + \sum_{i=1}^{d-1} g_i(f) + g_d(f) \\ &\geq g_0(f) + (d-1)g_1(f) + g_d(f) \\ &= 1 + (d-1)(f_{d-1} - d) + 1 \\ &= (d-1)f_{d-1} - (d+1)(d-2), \end{aligned}$$

as desired. For $j = 1, \ldots, d-2$ we have in a similar manner

$$\begin{aligned} f_j &= \sum_{i=0}^{d} \binom{i}{j} g_i(f) \\ &= \sum_{i=j}^{d-1} \binom{i}{j} g_i(f) + \binom{d}{j} g_d(f) \\ &\geq \sum_{i=j}^{d-1} \binom{i}{j} g_1(f) + \binom{d}{j} g_d(f) \\ &= (f_{d-1} - d) \sum_{i=j}^{d-1} \binom{i}{j} + \binom{d}{j}. \end{aligned}$$

Now, by Appendix 3, (9)

$$\begin{aligned} \sum_{i=j}^{d-1} \binom{i}{j} &= \sum_{k=0}^{d-1-j} \binom{j+k}{j} \\ &= \binom{j + (d-1-j) + 1}{d-1-j} = \binom{d}{j+1}, \end{aligned}$$

whence

$$\begin{aligned} f_j &\geq (f_{d-1} - d)\binom{d}{j+1} + \binom{d}{j} \\ &= \binom{d}{j+1} f_{d-1} - \binom{d+1}{j+1}(d-1-j), \end{aligned}$$

as desired. Hence, the Lower Bound Inequalities hold.

To deduce the Upper Bound Inequalities, we first prove that

$$(g_i(f) - g_{i-1}(f))^{\langle i \rangle} \leq \binom{f_{d-1} - d + i - 1}{i+1}, \qquad i = 1, \ldots, n-1. \tag{6}$$

For $i = 1$, we have

$$g_1(f) - g_0(f) = f_{d-1} - d - 1.$$

The 1-canonical representation of $f_{d-1} - d - 1$, of course, is given by

$$f_{d-1} - d - 1 = \binom{f_{d-1} - d - 1}{1}.$$

Hence,

$$(g_1(f) - g_0(f))^{\langle 1 \rangle} = \binom{f_{d-1} - d}{2},$$

proving (6) for $i = 1$. To prove (6) in general, we use induction. Suppose that (6) holds for i. Using (c) and the monotonicity of the pseudopower we then have

$$\begin{aligned}(g_{i+1}(f) - g_i(f))^{\langle i+1 \rangle} &\le ((g_i(f) - g_{i-1}(f))^{\langle i \rangle})^{\langle i+1 \rangle} \\ &\le \binom{f_{d-1} - d + i - 1}{i + 1}^{\langle i+1 \rangle}.\end{aligned}$$

To find the $(i + 1)$th pseudopower of

$$\binom{f_{d-1} - d + i - 1}{i + 1},$$

we need the $(i + 1)$-canonical representation. This, of course, is given by (!)

$$\binom{f_{d-1} - d + i - 1}{i + 1} = \binom{f_{d-1} - d + i - 1}{i + 1}.$$

Hence

$$\binom{f_{d-1} - d + i - 1}{i + 1}^{\langle i+1 \rangle} = \binom{f_{d-1} - d + i}{i + 2},$$

proving (6) for $i + 1$.

Using (6) we shall next prove that

$$g_i(f) \le \binom{f_{d-1} - d + i - 1}{i}, \qquad i = 0, \ldots, n. \tag{7}$$

This is certainly true for $i = 0, 1$, in fact with equality. We prove it in general by induction. Supposing that it holds for i, we have by (c) and (6),

$$\begin{aligned}g_{i+1}(f) &\le g_i(f) + (g_i(f) - g_{i-1}(f))^{\langle i \rangle} \\ &\le \binom{f_{d-1} - d + i - 1}{i} + \binom{f_{d-1} - d + i - 1}{i + 1} \\ &= \binom{f_{d-1} - d + i}{i + 1},\end{aligned}$$

as desired.

To complete the proof, note that using (1) and (a) we have

$$f_j = \sum_{i=0}^{n} \binom{i}{j} g_i(f) + \sum_{i=0}^{m} \binom{d-i}{j} g_i(f)$$

for $j = 0, \ldots, d$. Combining with (7), we get

$$f_j \leq \sum_{i=0}^{n} \binom{i}{j} \binom{f_{d-1} - d + i - 1}{i} + \sum_{i=0}^{m} \binom{d-i}{j} \binom{f_{d-1} - d + i - 1}{i}$$

for $j = 0, \ldots, d$, which is the desired inequality.

APPENDIX 1

Lattices

A relation $\preccurlyeq$ on a non-empty set M is called a *partial order* if it is reflexive, anti-symmetric and transitive, i.e. if

$$x \preccurlyeq x,$$

$$x \preccurlyeq y \wedge y \preccurlyeq x \Rightarrow x = y,$$

and

$$x \preccurlyeq y \wedge y \preccurlyeq z \Rightarrow x \preccurlyeq z$$

for all $x, y, z \in M$. A *partially ordered set* is a pair $(M, \preccurlyeq)$, where M is a non-empty set and $\preccurlyeq$ is a partial order on M.

In the following, let $(M, \preccurlyeq)$ be a partially ordered set, and let N be a subset of M. An element $x \in M$ is called a *lower bound* of N if $x \preccurlyeq y$ for all $y \in N$. Similarly, x is called an *upper bound* if $y \preccurlyeq x$ for all $y \in N$. An element $x \in M$ is called the *greatest lower bound* of N if x is a lower bound of N and $z \preccurlyeq x$ for any other lower bound z. The greatest lower bound of N is unique if it exists; it is denoted by inf N. Similarly, x is called the *least upper bound* of N if x is an upper bound of N and $x \preccurlyeq z$ for any other upper bound z. The least upper bound is unique if it exists; it is denoted by sup N.

A partially ordered set $(M, \preccurlyeq)$ is called a *lattice* if inf N and sup N exist for each non-empty finite subset N of M. If inf N and sup N exist for any subset N of M, then the lattice $(M, \preccurlyeq)$ is called a *complete lattice*.

Any finite lattice is complete.

If $(M, \preccurlyeq)$ is a partially ordered set such that inf N exists for all subsets N of M, then, in fact, $(M, \preccurlyeq)$ is a complete lattice. The same applies to sup N.

Let $(M, \preccurlyeq)$ be a lattice, and let M' be a non-empty subset of M. Then the partial order $\preccurlyeq$ on M induces a partial order on M' which we shall again denote by $\preccurlyeq$. We shall say that the partially ordered set $(M', \preccurlyeq)$ is a *sublattice* of the lattice $(M, \preccurlyeq)$ if $\inf N \in M'$ and $\sup N \in M'$ for each non-empty finite subset N of M'. (Here, of course, $\inf N$ and $\sup N$ mean the greatest lower bound and least upper bound, respectively, of N in $(M, \preccurlyeq)$.) Then $(M', \preccurlyeq)$ is also a lattice. Note that in the definition of a sublattice, we require more than just $(M', \preccurlyeq)$ being a lattice.

A mapping φ from one lattice $(M_1, \preccurlyeq)$ onto another lattice $(M_2, \preccurlyeq)$ is called an *isomorphism* when it is one-to-one and we have $x \preccurlyeq y$ if and only if $\varphi(x) \preccurlyeq \varphi(y)$ for all $x, y \in M_1$. If there exists an isomorphism from $(M_1, \preccurlyeq)$ onto $(M_2, \preccurlyeq)$, then we shall say that $(M_1, \preccurlyeq)$ and $(M_2, \preccurlyeq)$ are *isomorphic*. A mapping ψ from $(M_1, \preccurlyeq)$ onto $(M_2, \preccurlyeq)$ is called an *anti-isomorphism* when it is one-to-one and we have $x \preccurlyeq y$ if and only if $\psi(y) \preccurlyeq \psi(x)$ for all $x, y \in M_1$. If there exists an anti-isomorphism from $(M_1, \preccurlyeq)$ onto $(M_2, \preccurlyeq)$, then we shall say that $(M_1, \preccurlyeq)$ and $(M_2, \preccurlyeq)$ are *anti-isomorphic*.

Note that an isomorphism φ preserves inf and sup, i.e. $\varphi(\inf N) = \inf \varphi(N)$ and $\varphi(\sup N) = \sup \varphi(N)$, whereas an anti-isomorphism ψ reverses inf and sup, i.e., $\psi(\inf N) = \sup \psi(N)$ and $\psi(\sup N) = \inf \psi(N)$.

APPENDIX 2

Graphs

The intuitive picture of a (non-oriented) graph is that of a finite set of "vertices" and a finite set of "edges," each edge "joining" two distinct vertices and each two distinct vertices being joined by at most one edge. Formally, this may be expressed as follows: a *(non-oriented) graph* is a triple $\Gamma = (V, E, \gamma)$, where V (called the set of *vertices* of Γ) is a non-empty finite set, E (called the set of *edges* of Γ) is a set (necessarily finite), and γ (called the *incidence relation* of Γ) is a one-to-one mapping from E onto a subset of the set of all sets $\{x, y\}$ of two distinct elements of V.

When x and y are distinct vertices of a graph $\Gamma = (V, E, \gamma)$ and e is an edge of Γ such that $\gamma(e) = \{x, y\}$, then we shall say that *e joins x* and y, that x and y are the *endvertices* of e, that x and y are *incident* to e, and that e is *incident* to x and y. When x and y are distinct vertices of Γ joined by an edge, then we shall say that x and y are *adjacent*. The number of edges incident to a vertex x, i.e. the number of vertices adjacent to x, is called the *valence* of x.

A *path* in a graph Γ is a finite sequence of the form

$$x_1, e_1, x_2, e_2, \ldots, e_{n-1}, x_n,$$

where the x_i's are vertices of Γ, and the e_i's are edges of Γ such that each e_i joins x_i and x_{i+1}. The path is said to *join* x_1 and x_n, and x_1 and x_n are called the *endvertices* of the path. For technical reasons we allow $n = 1$, i.e. we allow trivial paths consisting of one vertex and no edges. A collection of paths joining two vertices x and y is called *independent* if x and y are the only vertices common to any two of the paths in the collection.

Two non-adjacent vertices x and y of a graph are said to be *separated* by a set W of vertices if every path joining x and y must contain a vertex of W.

A graph is said to be *connected* if any two distinct vertices can be joined by a path. A *disconnected* graph is one which is not connected. A graph is said to be *k-connected* (where k is a positive integer) if it has at least $k + 1$ vertices and any two distinct vertices can be joined by at least k independent paths. (Except for the trivial graphs with just one vertex, 1-connectedness is the same as connectedness.)

By a *subgraph* of a graph $\Gamma = (V, E, \gamma)$ we mean a graph $\Gamma' = (V', E', \gamma')$ such that $V' \subset V$, $E' \subset E$ and $\gamma'(e) = \gamma(e)$ for $e \in E'$. In general, each non-empty subset V' of the vertex set V of Γ is the vertex set of several subgraphs of Γ: each subset E' of the edge set E of Γ with the property that it only contains edges of Γ joining vertices in V' is the edge set of a subgraph with V' as the vertex set. If E' contains all the edges of Γ joining vertices in V', we call the resulting subgraph the subgraph *spanned* by V'.

Two non-adjacent vertices x and y of a graph $\Gamma = (V, E, \gamma)$ are separated by a set W of vertices (in the sense described above) if and only if the subgraph of Γ spanned by $V \setminus W$ is disconnected.

A path in a graph Γ may be considered as a subgraph. In general, it is not spanned by its set of vertices.

Let $\Gamma_i = (V_i, E_i, \gamma_i)$, $i = 1, \ldots, n$, be subgraphs of a graph $\Gamma = (V, E, \gamma)$. Let

$$V' := \bigcup_{i=1}^{n} V_i, \qquad E' := \bigcup_{i=1}^{n} E_i, \qquad \gamma'(e) := \gamma(e),\ e \in E'.$$

Then $\Gamma' = (V', E', \gamma')$ is a subgraph of Γ; we shall call it the *union* of the subgraphs Γ_i and denote it by $\Gamma_1 \cup \Gamma_2 \cup \cdots \cup \Gamma_n$.

In the main text we shall use the following two important connectedness results:

Theorem A2.1. *A graph $\Gamma = (V, E, \gamma)$ with at least $k + 1$ vertices is k-connected if and only if for each $k - 1$ vertices $x_1, \ldots, x_{k-1}$ of Γ, the subgraph Γ' of Γ spanned by*

$$V' := V \setminus \{x_1, \ldots, x_{k-1}\}$$

is connected.

Theorem A2.2. *Let Γ_1 and Γ_2 be k-connected subgraphs of a graph Γ. If Γ_1 and Γ_2 have at least k vertices in common, then their union $\Gamma_1 \cup \Gamma_2$ is also k-connected.*

The proofs of these two theorems will be given below. Theorem A2.2 is an easy consequence of Theorem A2.1. The main difficulty in proving Theorem A2.1 is taken care of by the following lemma:

Lemma A2.3. *Let x and y be non-adjacent vertices of a graph Γ. If the number of vertices of Γ needed to separate x and y equals k, then there are k independent paths in Γ joining x and y.*

Proof. It is trivial that the statement is true when $k = 1$. Suppose that it is not true for all k. Let k_0 be the smallest value of k for which the statement is not true. Let Γ_0 be a graph with the minimum number of vertices such that for appropriate non-adjacent vertices x_0 and y_0 of Γ_0, the number of vertices needed to separate x_0 and y_0 equals k_0, whereas the maximum number of independent paths joining x_0 and y_0 is at most $k_0 - 1$. By removing "superfluous" edges from Γ_0, if necessary, we may, in addition, assume that any graph Γ_0' obtained from Γ_0 by removing one edge has the property that only $k_0 - 1$ vertices are needed to separate x_0 and y_0 in Γ_0'.

We first prove:

(a) *No vertex of* Γ_0 *is adjacent to both* x_0 *and* y_0.

Suppose that a vertex v is adjacent to both x_0 and y_0. Let $\Gamma(v)$ denote the subgraph of Γ_0 spanned by all vertices of Γ_0 except v. Then clearly $k_0 - 1$ vertices are needed to separate x_0 and y_0 in $\Gamma(v)$. By the minimality property of k_0 we then see that there are $k_0 - 1$ independent paths in $\Gamma(v)$ joining x_0 and y_0. Along with the path whose vertices are x_0, v, y_0, this makes a total of k_0 independent paths in Γ_0 joining x_0 and y_0, a contradiction.

We next prove:

(b) *Let* W *be any set of* k_0 *vertices of* Γ_0 *separating* x_0 *and* y_0. *Then either every vertex in* W *is adjacent to* x_0, *or every vertex in* W *is adjacent to* y_0.

If for some $v \in W$, every path joining x_0 and y_0 passing through v contained at least one more vertex from W, then already $W \setminus \{v\}$ would separate x_0 and y_0, a contradiction. Therefore, for each $v \in W$ there is at least one path joining x_0 and y_0 such that v is the only vertex from W on that path. In particular, for each $v \in W$ there is a path joining x_0 and v such that v is the only vertex from W on the path. The union of all such paths is a subgraph of Γ_0. Adding to this subgraph the vertex y_0 plus k_0 "new" edges, each joining y_0 to a vertex in W, we obtain a new graph which we shall denote by $\Gamma(x_0)$. Changing the roles of x_0 and y_0, we obtain in a similar manner another new graph $\Gamma(y_0)$. Note that both $\Gamma(x_0)$ and $\Gamma(y_0)$ have the property that k_0 vertices are needed to separate x_0 and y_0; for any separating set in $\Gamma(x_0)$ or $\Gamma(y_0)$ must also be a separating set in Γ_0. Supposing that neither x_0 nor y_0 is adjacent to all vertices in W, it follows that both $\Gamma(x_0)$ and $\Gamma(y_0)$ have less vertices than Γ_0. By the minimality property of Γ_0 we then see that in both $\Gamma(x_0)$ and $\Gamma(y_0)$ there are k_0 independent paths joining x_0 and y_0. Removing from the paths in $\Gamma(x_0)$ the vertex y_0 and the (new) edge incident to y_0, we get k_0 paths in Γ_0, each joining x_0 to a vertex in W such that no vertex except x_0 belongs to more than one of the paths. In a similar way, removing from the paths in $\Gamma(y_0)$ the vertex x_0 and the (new) edge incident to x_0, we get k_0 paths in Γ_0, each joining y_0 to a vertex in W such that no vertex except y_0 belongs to more than one of the paths. Now, these $2k_0$ paths go together in pairs, each pair having some vertex from W in common. Each such pair determines a path joining x_0 and y_0. The resulting k_0 paths joining x_0 and y_0 are in fact independent. To see this, first note that each vertex in W belongs to exactly one of the paths. However, it is also impossible for a vertex not in W to belong to two of the paths. Suppose that z was such a common vertex of two paths p_1 and p_2. Then by the independence of the k_0 paths in $\Gamma(x_0)$ and the independence of the k_0 paths in $\Gamma(y_0)$, z had to lie between x_0 and a $v_1 \in W$ on one of the paths, say p_1, and between y_0 and a $v_2 \in W$ on the other path p_2. But then we could construct a path joining x_0 and y_0 via z not entering W, which is a contradiction.

To complete the proof of the lemma, let

$$x_0, e_0, u_1, e_1, u_2, \ldots, u_n, e_n, y_0 \tag{1}$$

be a path in Γ_0 joining x_0 and y_0. By the non-adjacency of x_0 and y_0, and (a), we must have $n \geq 2$. Let Γ_0' denote the subgraph of Γ_0 obtained by removing from Γ_0 the edge e_1. By one of our initial assumptions, only $k_0 - 1$ vertices are needed to separate x_0 and y_0 in Γ_0'. Let W' be such a separating set of vertices in Γ_0'. Then clearly both

$$W_1' := W' \cup \{u_1\}$$

and

$$W_2' := W' \cup \{u_2\}$$

separate x_0 and y_0 in Γ_0. It follows from (a) that u_1 is not adjacent to y_0 and u_2 is not adjacent to x_0. Application of (b) to W_1' then shows that each vertex in W' is adjacent to x_0, and application of (b) to W_2' shows that each vertex in W' is adjacent to y_0. Since the number of vertices in W' is $k_0 - 1$ which is at least 1, we get a contradiction by appealing to (a). This completes the proof of the lemma. □

PROOF (Theorem A2.1). Suppose first that Γ is k-connected. Let $x_1, \ldots, x_{k-1}$ be any $k - 1$ vertices of Γ, and let x and y be any two vertices from

$$V' := V \setminus \{x_1, \ldots, x_{k-1}\}.$$

By assumption, x and y can be joined by k independent paths in Γ. None of the vertices $x_1, \ldots, x_{k-1}$ belongs to more than one of these paths by the independence. Hence, at least one of the paths does not pass through any x_i. This shows that there is a path in the subgraph Γ' spanned by V' which joins x and y. In conclusion, Γ' is connected.

To prove the converse, let x and y be any two distinct vertices of Γ. If x and y are non-adjacent, it follows from the assumption that at least k vertices are needed to separate x and y. Lemma A2.3 next shows that there are at least k independent paths joining x and y, as desired. If x and y are adjacent, we argue as follows. Remove from Γ the edge joining x and y, and call the resulting graph Γ''. In Γ'', the vertices x and y are non-adjacent. Suppose that certain $k - 2$ vertices $x_1, \ldots, x_{k-2}$ would separate x and y in Γ''. In Γ'' there is at least one additional vertex z. Since x and y are separated, z must also be separated from at least one of the vertices, say y. But then the $k - 1$ vertices $x_1, \ldots, x_{k-2}, x$ separate y and z, which contradicts the assumption. Hence, in Γ'' at least $k - 1$ vertices are needed to separate x and y. Lemma A2.3 then tells that there are $k - 1$ independent paths in Γ'' joining x and y. Together with the path x, e, y, where e denotes the edge joining x and y in Γ, this makes k independent paths in Γ joining x and y. □

PROOF (Theorem A2.2). We use Theorem A2.1. Let $x_1, \ldots, x_{k-1}$ be any $k - 1$ vertices of $\Gamma_1 \cup \Gamma_2$, i.e. $x_1, \ldots, x_{k-1}$ belong to $V_1 \cup V_2$, where V_1 and V_2 denote the vertex set of Γ_1 and Γ_2, respectively. Let

$$V' := (V_1 \cup V_2)\setminus\{x_1, \ldots, x_{k-1}\},$$

and let Γ' denote the subgraph of $\Gamma_1 \cup \Gamma_2$ spanned by V'. Since at least k vertices are common to Γ_1 and Γ_2, at least one common vertex, say x_0, is distinct from all the x_i's, $i = 1, \ldots, k - 1$. Application of the "only if" part of Theorem A2.1 to Γ_1 shows that the subgraph Γ_1' of Γ_1 spanned by

$$V_1' := V_1\setminus\{x_1, \ldots, x_{k-1}\}$$

is connected. In the same manner, the subgraph Γ_2' of Γ_2 spanned by

$$V_2' := V_2\setminus\{x_1, \ldots, x_{k-1}\}$$

is connected. Since Γ_1' and Γ_2' have x_0 as a common vertex, it follows that $\Gamma_1' \cup \Gamma_2'$ is connected. Since $\Gamma_1' \cup \Gamma_2' = \Gamma'$, the desired conclusion follows from the "if" part of Theorem A2.1. □

Finally, we shall say a few words about oriented graphs. The intuitive picture of an oriented graph is that of a (non-oriented) graph as described above where, in addition, each edge is equipped with an "orientation". Formally, this may be stated as follows: an *oriented graph* is a triple $\Gamma = (V, E, \gamma)$, where V (called the set of *vertices* of Γ) is a non-empty finite set, E (called the set of *edges* of Γ) is a set (necessarily finite), and γ (called the *incidence relation* of Γ) is a one-to-one mapping from E onto a subset of the set of all ordered pairs (x, y) of two distinct elements of V; moreover, we require that if x and y are distinct vertices with $\gamma(e) = (x, y)$ for some edge e, then $\gamma(e') \neq (y, x)$ for all edges e'.

Each oriented graph $\Gamma = (V, E, \gamma)$ has an underlying non-oriented graph $\Gamma' = (V, E, \gamma')$, whereby $\gamma'(e) = \{x, y\}$ when $\gamma(e) = (x, y)$ or $\gamma(e) = (y, x)$. Therefore, everything that we have said above about non-oriented graphs also applies to oriented graphs, in the sense that it applies to the underlying non-oriented graph.

When $\Gamma = (V, E, \gamma)$ is an oriented graph, and $\gamma(e) = (x, y)$, then we say that the edge e is *oriented* towards y and away from x. The number of edges oriented towards x is called the *in-valence* of x, and the number of edges oriented away from x is called the *out-valence* of x. Hence, the sum of the in-valence of x and the out-valence of x equals the valence of x.

Each non-oriented graph may be turned into an oriented graph by choosing for each edge of the graph one of the two possible orientations of that edge. Formally, this means that if $\Gamma = (V, E, \gamma)$ is a non-oriented graph, then we get an oriented graph $\Gamma' = (V, E, \gamma')$ by choosing γ' such that $\gamma'(e) = (x, y)$ or $\gamma'(e) = (y, x)$ whenever $\gamma(e) = \{x, y\}$. Of course, γ' is not unique (unless Γ has no edges at all).

APPENDIX 3

Combinatorial Identities

In the main text we shall need certain identities involving binomial coefficients. The purpose of the present section is to give a unified exposition of these identities.

In the following, a, b and c always denote integers. Moreover, we always assume $b \geq 0$, whereas a and c may be negative.

Recall that

$$\binom{a}{b} := \frac{a(a-1)\cdots(a-b+1)}{b!}, \quad b \geq 1,$$

and

$$\binom{a}{0} := 1.$$

In particular,

$$\binom{a}{b} = \frac{a!}{b!(a-b)!}, \qquad b \leq a,$$

and

$$\binom{a}{b} = 0, \qquad 0 \leq a < b.$$

If $a \geq 0$, then

$$\binom{a}{b}$$

equals the number of choices of b elements among a elements.

We shall leave it to the reader to verify the following:

$$\binom{a}{b} = \binom{a}{a-b}. \tag{1}$$

$$\binom{a}{b} = (-1)^b \binom{-a+b-1}{b}. \tag{2}$$

$$\binom{a}{b} = (-1)^{a-b} \binom{-b-1}{a-b}. \tag{3}$$

$$\binom{a+1}{b+1} = \frac{a+1}{b+1}\binom{a}{b}. \tag{4}$$

$$\binom{a+1}{b+1} = \binom{a}{b+1} + \binom{a}{b}. \tag{5}$$

$$\binom{a}{b}\binom{c}{a} = \binom{c}{b}\binom{c-b}{a-b}. \tag{6}$$

(In (1), (3) and (6) it is understood that $a \geq b$.)

The combinatorial identities that we need in the main text are all consequences of the following basic identity, known as the *Vandermonde Convolution*:

$$\sum_{k=0}^{b} \binom{a}{k}\binom{c}{b-k} = \binom{a+c}{b}. \tag{7}$$

For $a, c \geq 0$, this is easy to prove. In fact,

$$\binom{a}{k}\binom{c}{b-k}$$

is the number of choices of b elements among $a + c$ elements such that k elements are chosen among certain a elements and the remaining $b - k$ elements are chosen among the remaining c elements; summing over k then yields (7). However, we need (7) for arbitrary integers a and c. We prove it by induction on b. For $b = 0$, it is trivial. Suppose that it holds for b. Then, using also (4), we have

$$\sum_{k=0}^{b+1}\binom{a}{k}\binom{c}{b+1-k}$$
$$=\sum_{k=0}^{b+1}\left(\frac{k}{b+1}+\frac{b+1-k}{b+1}\right)\binom{a}{k}\binom{c}{b+1-k}$$
$$=\sum_{k=1}^{b+1}\frac{k}{b+1}\binom{a}{k}\binom{c}{b+1-k}+\sum_{k=0}^{b}\frac{b+1-k}{b+1}\binom{a}{k}\binom{c}{b+1-k}$$
$$=\sum_{h=0}^{b}\frac{h+1}{b+1}\binom{a}{h+1}\binom{c}{b+1-(h+1)}+\sum_{k=0}^{b}\frac{b+1-k}{b+1}\binom{a}{k}\binom{c}{b+1-k}$$
$$=\sum_{h=0}^{b}\frac{a}{b+1}\binom{a-1}{h}\binom{c}{b-h}+\sum_{k=0}^{b}\frac{c}{b+1}\binom{a}{k}\binom{c-1}{b-k}$$
$$=\frac{a}{b+1}\binom{(a-1)+c}{b}+\frac{c}{b+1}\binom{a+(c-1)}{b}$$
$$=\frac{a+c}{b+1}\binom{a+c-1}{b}$$
$$=\binom{a+c}{b+1}$$

Hence, (7) holds for $b+1$, as desired.

Taking $c=-1$ in (7) and using the fact that by definition

$$\binom{-1}{b-k}=(-1)^{b-k},$$

we get

$$\sum_{k=0}^{b}(-1)^k\binom{a}{k}=(-1)^b\binom{a-1}{b}$$
$$=\binom{b-a}{b}. \tag{8}$$

Since

$$\binom{a+k}{k}=(-1)^k\binom{-a-1}{k},$$

cf. (2), we see that (8) is equivalent to

$$\sum_{k=0}^{b}\binom{a+k}{k}=(-1)^b\binom{-a-2}{b}$$
$$=\binom{a+b+1}{b}. \tag{9}$$

We shall next prove:

$$\sum_{k=0}^{b}(-1)^k\binom{k}{a}\binom{c}{k} = (-1)^b\binom{c}{a}\binom{c-a-1}{b-a}, \qquad 0 \le a \le b. \tag{10}$$

The proof of this uses (6) and (8). We have

$$\begin{aligned}\sum_{k=0}^{b}(-1)^k\binom{k}{a}\binom{c}{k} &= \sum_{k=a}^{b}(-1)^k\binom{k}{a}\binom{c}{k}\\ &= \sum_{k=a}^{b}(-1)^k\binom{c}{a}\binom{c-a}{k-a}\\ &= (-1)^a\binom{c}{a}\sum_{h=0}^{b-a}(-1)^h\binom{c-a}{h}\\ &= (-1)^a\binom{c}{a}(-1)^{b-a}\binom{c-a-1}{b-a}\\ &= (-1)^b\binom{c}{a}\binom{c-a-1}{b-a},\end{aligned}$$

completing the proof of (10).

A particular case of (10) is the following:

$$\sum_{k=0}^{b}(-1)^k\binom{k}{a}\binom{c}{k} = (-1)^a\delta(a,c), \qquad 0 \le a \le b, \quad 0 \le c \le b. \tag{11}$$

In fact, if $a = c$, then

$$\begin{aligned}(-1)^b\binom{c}{a}\binom{c-a-1}{b-a} &= (-1)^b \cdot 1 \cdot \binom{-1}{b-a}\\ &= (-1)^b(-1)^{b-a}\\ &= (-1)^a;\end{aligned}$$

if $a > c$, then

$$\binom{c}{a} = 0;$$

and if $a < c$, then $0 \le c - a - 1 < b - a$, whence

$$\binom{c-a-1}{b-a} = 0.$$

This completes the proof of (11).

Our final identity is the following:

$$\sum_{k=0}^{b}(-1)^k\binom{k}{a}\binom{c}{b-k} = (-1)^b\binom{b-c}{b-a}, \qquad 0 \le a \le b. \tag{12}$$

Using (3), (7) and (2), we have

$$\begin{aligned}
\sum_{k=0}^{b}(-1)^k\binom{k}{a}\binom{c}{b-k} &= \sum_{k=a}^{b}(-1)^k\binom{k}{a}\binom{c}{b-k}\\
&= \sum_{k=a}^{b}(-1)^k(-1)^{k-a}\binom{-a-1}{k-a}\binom{c}{b-k}\\
&= (-1)^a\sum_{h=0}^{b-a}\binom{-a-1}{h}\binom{c}{(b-a)-h}\\
&= (-1)^a\binom{(-a-1)+c}{b-a}\\
&= (-1)^a(-1)^{b-a}\binom{-(-a-1)-c+(b-a)-1}{b-a}\\
&= (-1)^b\binom{b-c}{b-a},
\end{aligned}$$

completing the proof of (12).

The combinatorial identities of this section may be interpreted as statements about products of matrices. As an important example, let us consider the identity (11). Let B and D denote the $(b+1)\times(b+1)$ matrices

$$B := \left((-1)^{i+j}\binom{j}{i}\right)_{i=0,\dots,b;\, j=0,\dots,b},$$

$$D := \left(\binom{j}{i}\right)_{i=0,\dots,b;\, j=0,\dots,b}.$$

Then the identity (11) tells that B and D are mutually inverse matrices.

Bibliographical Comments

Modern convexity theory was founded around the beginning of this century by H. Minkowski, C. Carathéodory and others. Most of Chapter 1 goes back to that period. As general references to the topics discussed in Chapter 1 we recommend the books of T. Bonnesen and W. Fenchel [11], R. T. Rockafellar [21] and B. Grünbaum [12].

The contents of Sections 7–12 in Chapter 2 are also classical. The cyclic polytopes of Section 13 were already discussed by Carathéodory; he also discovered their neighbourliness, cf. Section 14. Around 1955, these polytopes were rediscovered by D. Gale. The main result of Section 15, namely Theorem 15.6, is due to M. Balinski (1961). As a general reference to Chapter 2, Grünbaum's book is indespensable. See also the books of Rockafellar (on topics related to Sections 7–9), J. Bair and R. Fourneau [1] and P. McMullen and G. C. Shephard [18].

In Chapter 3, Euler's Relation was discovered by L. Euler in 1752 for the case $d = 3$. For the interesting history of Euler's Relation, see [12, Section 8.6]. Our exposition follows [18].

For the history of the Dehn–Sommerville Relations, see [12, Section 9.8]. Our Theorems 17.1, 17.5 and 17.6 appear in their dual forms in [12] on pp. 146, 148 and 161, respectively.

The Upper Bound Theorem was conjectured by T. S. Motzkin in 1957. In the following years, a number of particular cases were settled by V. Klee and others, until the final proof was given by P. McMullen [16]. Both Motzkin's conjecture and McMullen's proof (and the intermediate papers as well) were formulated in the setting of simplicial polytopes. Our exposition, using simple polytopes, follows A. Bondesen and A. Brøndsted [10]. For more details about the history of the Upper Bound Theorem up to 1967, see [12, Section 10.1].

The Lower Bound Theorem was proved by D. Barnette [2, 3]. Our exposition is based on [2, 3] and V. Klee [14]. For the history of the Lower Bound Theorem up to 1967, see [12, Section 10.2]. The fact that for $d \geq 4$, the truncation polytopes are the only polytopes for which equality holds was proved by L. J. Billera and C. W. Lee [5].

Upper and lower bound inequalities for simple unbounded polyhedral sets have been obtained by V. Klee [13], A. Björner [7], L. J. Billera and C. W. Lee [6] and C. W. Lee [15].

Theorem 20.1 was conjectured by P. McMullen [17] (in the setting of simplicial polytopes). He also proved the theorem for certain particular cases and showed that the conjecture would imply the Upper Bound Theorem. Another paper (preceding [17]), related to Theorem 20.1, is by P. McMullen and D. W. Walkup [19]; here the necessity of condition (b) of Theorem 20.1 is conjectured and it shown that (a) and (b) imply the Lower Bound Theorem.

The sufficiency of McMullen's conditions was established by L. J. Billera and C. W. Lee [4, 5]. In their proof (which is formulated in the setting of simplicial polytopes) they produce a simplicial d-polytope with a given $f = (f_0, \ldots, f_{d-1})$ as its f-vector by taking the vertex-figure at z of a $(d + 1)$-polytope of the form $\operatorname{conv}(Q \cup \{z\})$, where Q is a suitably chosen cyclic $(d + 1)$-polytope and z is a suitably chosen point outside Q.

The necessity of McMullen's conditions was established by R. P. Stanley [22]. Stanley's proof (which is formulated in the setting of simplicial polytopes) uses advanced algebraic geometry; it would be very desirable to have a more elementary proof.

A conjecture on the characterization of f-vectors of simple unbounded polyhedral sets has been formulated by L. J. Billera and C. W. Lee [6].

It has been conjectured that the f-vectors of simplicial (or simple) d-polytopes P are *unimodal*, i.e.

$$f_0(P) \leq f_1(P) \leq \cdots \leq f_k(P) \geq f_{k+1}(P) \geq \cdots \geq f_{d-1}(P)$$

for some k. (The necessity of (a) and (b) in Theorem 20.1 shows that $(g_0(P), \ldots, g_d(P))$ is unimodal for every simple d-polytope P.) According to A. Björner [8], it can be shown that for any simplicial d-polytope P,

$$f_0(P) < f_1(P) < \cdots < f_{\lfloor d/2 \rfloor - 1}(P) \leq f_{\lfloor d/2 \rfloor}(P),$$

$$f_{\lfloor 3(d-1)/4 \rfloor}(P) > \cdots > f_{d-2}(P) > f_{d-1}(P).$$

These inequalities immediately imply unimodality for $d \leq 8$. It is even possible to show unimodality for all $d \leq 15$ by checking each d separately, cf. [8]. But unimodality does not hold in general: one knows 20-dimensional simplicial polytopes P (with on the order of 10^{13} vertices) such that $f_{11}(P) > f_{12}(P) < f_{13}(P)$, cf. [5], [8].

To conclude the comments on Chapters 1–3, let us mention, without going into detail, that many combinatorial results about convex polytopes admit extensions to more general geometric objects.

Our exposition of Menger's Theorem (Theorem A2.1 of Appendix 2) is based on the book of B. Bollobás [9] which we also recommend as a general reference to graph theory. As a general reference to combinatorial identities (Appendix 3) we recommend the book of J. Riordan [20].

Bibliography

1. Bair, J. and Fourneau, R. 1980. *Etude Géometrique des Espaces Vectoriels, II. Polyèdres et Polytopes Convexes*. Lecture Notes in Mathematics, No. 802. New York–Heidelberg–Berlin: Springer-Verlag.
2. Barnette, D. 1971. The minimum number of vertices of a simple polytope. *Israel J. Math.*, **10**, 121–125.
3. Barnette, D. 1973. A proof of the lower bound conjecture for convex polytopes. *Pacific J. Math.*, **46**, 349–354.
4. Billera, L. J. and Lee, C. W. 1980. Sufficiency of McMullen's conditions for f-vectors of simplicial polytopes. *Bull. Amer. Math. Soc.* (*N.S.*), **2**, 181–185.
5. Billera, L. J. and Lee, C. W. 1981. A proof of the sufficiency of McMullen's conditions for f-vectors of simplicial convex polytopes. *J. Combin. Theory, Ser. A*, **31**, 237–255.
6. Billera, L. J. and Lee, C. W. 1981. The number of faces of polytope pairs and unbounded polyhedra. *European J. Combin.*, **2**, 307–322.
7. Björner, A. 1980. The minimum number of faces of a simple polyhedron. *European J. Combin.*, **1**, 27–31.
8. Björner, A. 1981. The unimodality conjecture for convex polytopes. *Bull. Amer. Math. Soc.* (*N.S.*), **4**, 187–188.
9. Bollobás, B. 1979. *Graph Theory—An Introductory Course*. Graduate Texts in Mathematics, Vol. 63. New York–Heidelberg–Berlin: Springer-Verlag.
10. Bondesen, A. and Brøndsted, A. 1980. A dual proof of the upper bound conjecture for convex polytopes. *Math. Scand.*, **46**, 95–102.
11. Bonnesen, T. and Fenchel, W. 1934. *Theorie der Konvexen Körper*. Ergebnisse der Mathematik und ihrer Grenzgebiete, Bd. 3, Nr. 1. Berlin: Verlag von Julius Springer. (Berichtigter Reprint 1974. New York–Heidelberg–Berlin: Springer-Verlag.)
12. Grünbaum, B. 1967. *Convex Polytopes*. London–New York–Sydney: Wiley.
13. Klee, V. 1974. Polytope pairs and their relationship to linear programming. *Acta Math.*, **133**, 1–25.
14. Klee, V. 1975. A d-pseudomanifold with f_0 vertices has at least $df_0 - (d-1)(d+2)$ d-simplices. *Houston J. Math.*, **1**, 81–86.
15. Lee, C. W. 1981. Bounding the numbers of faces of polytope pairs and simple polyhedra. Research report RC 8774, IBM Thomas J. Watson Research Center, Yorktown Heights, NY.

16. McMullen, P. 1971. The maximum numbers of faces of a convex polytope. *Mathematika*, **17**, 179–184.
17. McMullen, P. 1971. The numbers of faces of simplicial polytopes. *Israel J. Math.*, **9**, 559–570.
18. McMullen, P. and Shephard, G. C. 1971. *Convex Polytopes and the Upper Bound Conjecture*. London Mathematical Society Lecture Note Series, Vol. 3. Cambridge: Cambridge University Press.
19. McMullen, P. and Walkup, D. W. 1971. A generalized lower-bound conjecture for simplicial polytopes. *Mathematika*, **18**, 264–273.
20. Riordan, J. 1968. *Combinatorial Identities*. New York–London–Sydney: Wiley.
21. Rockafellar, R. T. 1970. *Convex Analysis*. Princeton, N. J.: Princeton University Press.
22. Stanley, R. P. 1980. The number of faces of a simplicial convex polytope. *Adv. Math.*, **35**, 236–238.

List of Symbols

$\emptyset$	empty set
$\mathbb{N}$	set of natural numbers
$\mathbb{R}$	set of real numbers
$\mathbb{R}^d$	set of d-tuples $(\alpha_1, \ldots, \alpha_d)$ of real numbers
$[0, 1]$	set of real numbers a with $0 \leq a \leq 1$
$]0, 1]$	set of real numbers a with $0 < a \leq 1$
$[0, 1[$	set of real numbers a with $0 \leq a < 1$
$]0, 1[$	set of real numbers a with $0 < a < 1$
o	zero element $(0, \ldots, 0)$ in $\mathbb{R}^d$
aff M	affine hull of set M 7
bd M	boundary of set M
card M	cardinality of set M
cl M	closure of set M
clconv M	closed convex hull of set M 24
conv M	convex hull of set M 12
dim A	dimension of affine subspace A 7
dim C	dimension of convex set C 19
dim L	dimension of linear subspace L 5
exp C	set of exposed points of closed convex set C 31
ext C	set of extreme points of closed convex set C 30
int M	interior of set M
rb C	relative boundary of convex set C 19
ri C	relative interior of convex set C 19
span M	linear hull of set M 5

$\mathcal{M}_d$	moment curve in $\mathbb{R}^d$ 85
$\mathcal{P}^d$	set of d-polytopes 98
$\mathcal{P}^d_\sigma$	set of simple d-polytopes 104
$\mathcal{P}^d_s$	set of simplicial d-polytopes 104
$\mathcal{E}(C)$	set of exposed faces of closed convex set C 31
$\mathcal{F}(C)$	set of faces of closed convex set C 30
$\mathcal{F}(P/x_0)$	set of faces F of polytope P with $x_0 \in F \subset P$ 68
$\mathcal{F}(F_2/F_1)$	set of faces F of polytope P with $F_1 \subset F \subset F_2$ 69
$\mathcal{F}_j(P)$	set of j-faces of polytope P 100
$\mathcal{G}(P)$	graph of polytope P 93
$\mathcal{G}(P, w)$	oriented graph of polytope P determined by admissible vector w 94
$\mathcal{G}(\mathcal{S})$	graph of facet system $\mathcal{S}$ 96
$f(P)$	f-vector of polytope P 66
$f(\mathcal{P}^d)$	set of f-vectors of d-polytopes 98
$f(\mathcal{P}^d_\sigma)$	set of f-vectors of simple d-polytopes 104
$f_j(P)$	number of j-faces of polytope P 66
$f_j(F_2/F_1)$	number of j-faces F of polytope P with $F_1 \subset F \subset F_2$ 103
$g_i(P)$	number of vertices in $\mathcal{G}(P, w)$ with in-valence i 114
$g_i(f)$	130
$\Phi_j(d, p)$	113
$\varphi_j(d, p)$	121
$B(z, r)$	closed ball centred at z with radius r 38
$C(n, d)$	type of cyclic polytope 85
$H(y, \alpha)$	9
$H(\varepsilon, 1 - (-1)^d)$	Euler Hyperplane in $\mathbb{R}^d$ 102
$K(y, \alpha)$	9
$\lVert u \rVert$	standard Euclidean norm in $\mathbb{R}^d$
$\lfloor a \rfloor$	greatest integer $\leq a$
$M^\perp$	annihilator of set M, orthogonal complement of span M
M°	polar of set M 37
$M^{\circ\circ}$	bipolar of set M, polar of M° 38
$F^\triangle$	exposed face conjugate to exposed face F 40
$F^{\triangle\triangle}$	exposed face conjugate to exposed face $F^\triangle$ 41
$\langle u, v \rangle$	standard inner product in $\mathbb{R}^d$
$[x_1, x_2]$	closed segment 11
$]x_1, x_2]$	half-open segment 11
$[x_1, x_2[$	half-open segment 11
$]x_1, x_2[$	open segment 11
$h^{\langle k \rangle}$	kth pseudopower of h 130

$\sum_{i=1}^{n}{}^{a}\, \lambda_i x_i$	affine combination 7
$\sum_{i=1}^{n}{}^{c}\, \lambda_i x_i$	convex combination 11
m	*in Sections* 17–20: $\lfloor (d-1)/2 \rfloor$
n	*in Sections* 17–20: $\lfloor d/2 \rfloor$
$\subset$	subset of
$\subsetneqq$	proper subset of
$\not\subset$	not subset of

Index

Graduate Texts in Mathematics

1 TAKEUTI/ZARING. Introduction to Axiomatic Set Theory. 2nd ed.
2 OXTOBY. Measure and Category. 2nd ed.
3 SCHAEFFER. Topological Vector Spaces.
4 HILTON/STAMMBACH. A Course in Homological Algebra. 3rd printing.
5 MACLANE. Categories for the Working Mathematician.
6 HUGHES/PIPER. Projective Planes.
7 SERRE. A Course in Arithmetic.
8 TAKEUTI/ZARING. Axiometic Set Theory.
9 HUMPHREYS. Introduction to Lie Algebras and Representation Theory. 3rd printing, revised.
10 COHEN. A Course in Simple Homotopy Theory.
11 CONWAY. Functions of One Complex Variable. 2nd ed.
12 BEALS. Advanced Mathematical Analysis.
13 ANDERSON/FULLER. Rings and Categories of Modules.
14 GOLUBITSKY/GUILLEMIN. Stable Mappings and Their Singularities.
15 BERBERIAN. Lectures in Functional Analysis and Operator Theory.
16 WINTER. The Structure of Fields.
17 ROSENBLATT. Random Processes. 2nd ed.
18 HALMOS. Measure Theory.
19 HALMOS. A Hilbert Space Problem Book. 2nd ed., revised.
20 HUSEMOLLER. Fibre Bundles. 2nd ed.
21 HUMPHREYS. Linear Algebraic Groups.
22 BARNES/MACK. An Algebraic Introduction to Mathematical Logic.
23 GREUB. Linear Algebra. 4th ed.
24 HOLMES. Geometric Functional Analysis and its Applications.
25 HEWITT/STROMBERG. Real and Abstract Analysis. 5th printing.
26 MANES. Algebraic Theories.
27 KELLEY. General Topology.
28 ZARISKI/SAMUEL. Commutative Algebra. Vol. I.
29 ZARISKI/SAMUEL. Commutative Algebra. Vol. II.
30 JACOBSON. Lectures in Abstract Algebra I: Basic Concepts.
31 JACOBSON. Lectures in Abstract Algebra II: Linear Algebra.
32 JACOBSON. Lectures in Abstract Algebra III: Theory of Fields and Galois Theory.
33 HIRSCH. Differential Topology.
34 SPITZER. Principles of Random Walk. 2nd ed.
35 WERMER. Banach Algebras and Several Complex Variables. 2nd ed.
36 KELLEY/NAMIOKA et al. Linear Topological Spaces.
37 MONK. Mathematical Logic.
38 GRAUERT/FRITZSCHE. Several Complex Variables.
39 ARVESON. An Invitation to C^*-Algebras.
40 KEMENY/SNELL/KNAPP. Denumerable Markov Chains. 2nd ed.
41 APOSTOL. Modular Functions and Dirichlet Series in Number Theory.
42 SERRE. Linear Representations of Finite Groups.
43 GILLMAN/JERISON. Rings of Continuous Functions.
44 KENDIG. Elementary Algebraic Geometry.
45 LOÈVE. Probability Theory I. 4th ed.
46 LOÈVE. Probability Theory II. 4th ed.

47 MOISE. Geometric Topology in Dimensions 2 and 3.
48 SACHS/WU. General Relativity for Mathematicians.
49 GRUENBERG/WEIR. Linear Geometry. 2nd ed.
50 EDWARDS. Fermat's Last Theorem. 2nd printing.
51 KLINGENBERG. A Course in Differential Geometry.
52 HARTSHORNE. Algebraic Geometry.
53 MANIN. A Course in Mathematical Logic.
54 GRAVER/WATKINS. Combinatorics with Emphasis on the Theory of Graphs.
55 BROWN/PEARCY. Introduction to Operator Theory I: Elements of Functional Analysis.
56 MASSEY. Algebraic Topology: An Introduction.
57 CROWELL/FOX. Introduction to Knot Theory.
58 KOBLITZ. p-adic Numbers, p-adic Analysis, and Zeta-Functions.
59 LANG. Cyclotomic Fields.
60 ARNOLD. Mathematical Methods in Classical Mechanics. 3rd printing.
61 WHITEHEAD. Elements of Homotopy Theory.
62 KARGAPOLOV/MERZLJAKOV. Fundamentals of the Theory of Groups.
63 BOLLOBÁS. Graph Theory.
64 EDWARDS. Fourier Series. Vol. I. 2nd ed.
65 WELLS. Differential Analysis on Complex Manifolds. 2nd ed.
66 WATERHOUSE. Introduction to Affine Group Schemes.
67 SERRE. Local Fields.
68 WEIDMANN. Linear Operators in Hilbert Spaces.
69 LANG. Cyclotomic Fields II.
70 MASSEY. Singular Homology Theory.
71 FARKAS/KRA. Riemann Surfaces.
72 STILLWELL. Classical Topology and Combinatorial Group Theory.
73 HUNGERFORD. Algebra.
74 DAVENPORT. Multiplicative Number Theory. 2nd ed.
75 HOCHSCHILD. Basic Theory of Algebraic Groups and Lie Algebras.
76 IITAKA. Algebraic Geometry.
77 HECKE. Lectures on the Theory of Algebraic Numbers.
78 BURRIS/SANKAPPANAVAR. A Course in Universal Algebra.
79 WALTERS. An Introduction to Ergodic Theory.
80 ROBINSON. A Course in the Theory of Groups.
81 FORSTER. Lectures on Riemann Surfaces.
82 BOTT/TU. Differential Forms in Algebraic Topology.
83 WASHINGTON. Introduction to Cyclotomic Fields.
84 IRELAND/ROSEN. A Classical Introduction Modern Number Theory.
85 EDWARDS. Fourier Series. Vol. II. 2nd ed.
86 VAN LINT. Introduction to Coding Theory.
87 BROWN. Cohomology of Groups.
88 PIERCE. Associative Algebras.
89 LANG. Introduction to Algebraic and Abelian Functions. 2nd ed.
90 BRØNDSTED. An Introduction to Convex Polytopes.

9 780387 907222